# JUST COOL IT!

# DAVID SUZUKI
## & IAN HANINGTON

# JUST COOL IT!

A POST–PARIS
AGREEMENT
GAME PLAN

## The Climate Crisis and What We Can Do

DAVID SUZUKI INSTITUTE

GREYSTONE BOOKS
Vancouver/Berkeley

Greystone Books Ltd.
www.greystonebooks.com

David Suzuki Institute
219–2211 West 4th Avenue
Vancouver BC Canada V6K 4S2
www.davidsuzuki.org

Cataloguing data available from Library and Archives Canada
ISBN 978-1-77164-259-0 (pbk.)
ISBN 978-1-77164-260-6 (epub)

Editing by Shirarose Wilensky
Proofreading by Stefania Alexandru
Cover design by Naomi MacDougall
Text design by Nayeli Jimenez
Cover photograph by shutterstock.com
Printed and bound in Canada on ancient-forest-friendly paper by Friesens

We gratefully acknowledge the support of the Canada Council for the Arts, the British Columbia Arts Council, the Province of British Columbia through the Book Publishing Tax Credit, and the Government of Canada for our publishing activities.

Canadä

There is a tide in the affairs of men.
Which, taken at the flood, leads on to fortune;
Omitted, all the voyage of their life
Is bound in shallows and in miseries.
On such a full sea are we now afloat.

—Shakespeare, *Julius Caesar*

# Contents

**I FIRST LEARNED OF** global warming sometime in the mid-1970s but thought of it as a slow-motion catastrophe that we had lots of time to work on. Meanwhile, at home in British Columbia, I was focused on pressing issues such as clear-cut logging, toxic pollution, and overfishing. In 1988, I visited Australia as a guest of the newly formed government think tank Commission for the Future, where I was shown the data climatologists had gathered. I realized climate change demanded immediate action, because although greenhouse gas accumulation and changes in global temperature and climate went unnoticed by most people, scientists were confirming that emissions and temperatures were rising at unprecedented speed.

In 1988, public concern about the environment had risen to such a level that George H.W. Bush promised that if Americans elected him, he would be an "environmental president." Once in power, he revealed his strong support for fossil fuels, describing pipelines as a boon to caribou and refusing to attend the 1992 Earth Summit in Rio de Janeiro unless the proposed international climate agreement was watered down. He attended after the emissions target was set at stabilization of

1990 levels by 2000. (Over and over, we see politicians making commitments to targets that will only be reached long after they are out of office.)

In 1988, Canadian prime minister Brian Mulroney asked politician and diplomat Stephen Lewis to chair some sessions for a major climate conference in Toronto. Lewis told me attendees were so concerned that they put out a news release at the conference's end warning that global warming represented a threat to human survival second only to nuclear war and calling for a 20 percent reduction in greenhouse gas emissions over fifteen years. Subsequently, numerous studies in Australia, Canada, Sweden, and the U.S. concluded that the target was readily achievable and would result in net savings that far exceeded costs. The problem was that few politicians were willing to take the heat for an initially costly program when someone else would get credit for meeting the target while saving large amounts of money.

In 1989, I hosted a five-part CBC radio series, *It's a Matter of Survival*, describing the devastation that would occur if we were to carry on with business as usual. As a result, the pre-email audience sent in more than seventeen thousand letters, most asking about solutions. In response, my wife, Tara, and I, along with a few other activists, established the David Suzuki Foundation in 1990. Our goal was to use the best scientific information to seek the underlying causes of our destructiveness and to find solutions that would protect nature and move us onto a sustainable path. We were impelled by the urgency of the message in *It's a Matter of Survival* and the World-watch Institute's designation of the 1990s as the turnaround decade, a ten-year time frame for humanity to shift onto a better path.

The CBC TV show I host, The Nature of Things, presented its first program on global warming in 1989, with scientists and politicians calling for action to reduce the threat of climate change. Ever since, human-induced climate change and its global consequences have been included in dozens of programs on The Nature of Things.

In 1988, the World Meteorological Organization and the United Nations Environment Programme set up the Intergovernmental Panel on Climate Change, or IPCC, a scientific body charged with examining the state of the climate and the political and economic implications. Although the IPCC was to be science-based and objective, it was subjected to enormous pressure from the fossil fuel industry and oil producing nations, so its reports and conclusions were authoritative but extremely cautious and conservative. In 1995, I attended the IPCC meeting in Geneva, where evidence was provided that the human imprint on climate change was discernible and action to reduce greenhouse gas emissions was necessary—a conclusion most climatologists had reached years earlier.

In 1997, world leaders gathered in Kyoto, Japan, to discuss ways to reduce emissions. Although some countries tried to block or stall action, delegates agreed on targets to be achieved by 2012. Because evidence showed economic growth and fossil fuel exploitation in industrialized nations were the primary causes of the climate crisis, those nations would be required to cap and reduce emissions while countries in the developing world could grow their economies without restricting fossil fuel use. Industrialized countries were given a target of reducing emissions 5 to 6 percent below 1990 levels by 2012, after which all nations would agree to a comprehensive reduction plan.

As soon as the Kyoto conference ended, critics began to denounce the agreement, saying it was unfair that countries such as China and India were exempt, that the evidence for the human role in climate change wasn't compelling, that meeting the target would be expensive and a threat to economies, and that individual nations should determine their own solutions. These were all bogus arguments driven by conservatives and the fossil fuel industry, but they succeeded in diminishing the fervour to reach targets. Little was done subsequently to seriously reduce emissions. If we in the rich countries, who had created the problem in the first place with our carbon-intensive economies, were not willing to reduce emissions, India, China, and all the other developing countries could not be expected to curb theirs. Meanwhile, each IPCC report provided greater certainty about the rising threat of climate change and the human contribution to it. The 2013–14 *Fifth Assessment Report* was the most conclusive and led to the 2015 Paris Agreement, signed by 195 nations responsible for 95 percent of the world's emissions. It was the first universal accord to spell out ways to confront climate change, requiring industrialized nations to transition from fossil fuels to 100 percent renewable energy by 2050, and developing nations by about 2080.

I'm from Canada, an industrialized country especially vulnerable to the effects of climate change. We're a northern nation, and the impacts are greater in polar areas. For decades, Inuit have sounded the alarm about rapidly rising temperatures affecting ice formation and loss, and the enormous ecological ramifications. Canada also has the longest marine coastline in the world, and sea level rise from thermal expansion of water alone will have a huge impact on our coastal areas. As the vast ice sheets of Greenland and Antarctica melt or slide into the oceans, sea levels will rise by yards and Canada's boundaries will undergo radical change. Glaciers are receding with shocking speed, and people and other species are already feeling the effects on Canada's vast network of rivers and lakes. The economic impact on climate-sensitive activities—agriculture, forestry, fisheries, tourism, winter sports—is already being felt and will worsen if our emissions are not rapidly reduced.

Across Canada, I have interviewed local politicians and people who still spend time outside—farmers, hunters, fishers, loggers, hydrologists, skiers—all of whom confirm the reality that the climate is changing.

I have also learned from personal experience that the changes are real. Thirty-seven years ago, when my daughter Severn was born, Tara and I wanted to be sure she understood that food is seasonal and should be celebrated when it's available. So we chose cherries. Each year, at the end of June, we set off camping, fishing, and working our way toward the Okanagan Valley, where we would gorge on cherries and pick boxes of the fruit to send to friends. It became a wonderful ritual. Severn was joined by a sister, Sarika, and as they became teenagers, boyfriends would accompany us and show off with their ability to climb and pick. Then grandchildren began to join us. In 2015, I called the organic farmers in Oliver to ask them to save trees for us to pick when we arrived on July 1. "Sorry," they told me, "the cherries are already finished." They had ripened three weeks early! In 2016, we went in mid-June for the first time and cherry season was already well underway. Farmers know that climate change is real.

Five or six years ago, I flew north to a meeting in Smithers, B.C. I was stunned to look out on both sides of the plane to see that the forest had turned bright red! Pine trees were dying because of an explosion in mountain pine beetles, an indigenous insect the size of a grain of rice, no longer controlled by cold winter temperatures. Immense clouds of beetles attacked billions of dollars' worth of pine trees in an unprecedented epidemic that has blown across the mountains into Alberta, the western portion of the halo of boreal pines that extends all the way to the East Coast. Indigenous people of the north, as well as loggers, hunters, and outfitters, will tell you climate change has kicked in.

The debate over the reality or rate of a changing climate is over. It's real, it's happening far more rapidly than we expected, and it's time for us to act. This crisis, exacerbated by the failure to respond to the challenge in 1988, is now a huge opportunity. I was a student in the U.S. when the Soviet Union launched *Sputnik* in 1957. In the ensuing months and years, American rockets kept blowing up while the Russians demonstrated their superiority by launching into space an animal, a man, a team, and a woman, all before the Americans. I watched with admiration as the United States responded without hesitation or concerns about cost. In 1961, when President John F. Kennedy announced the plan to get American astronauts to the moon and back within a decade, it wasn't at all clear how the target would be achieved. The important thing was the commitment to beat the Soviet Union. With a lot of money, energy, and creativity, the U.S. became the only nation to land people on the moon and to realize spinoffs that no one anticipated: GPS, twenty-four-hour news channels, laptop computers, space blankets, ear thermometers, and more. To this day, when Nobel science prizes are announced, Americans win a disproportionate share, in large part because the country committed to beating the Russians to the moon more than fifty years ago. History informs us once we make the commitment to beat the challenge of climate change, all sorts of unexpected benefits will result.

This book aims to provide a comprehensive view of global warming: the science and its history, the consequences, the obstacles to overcoming the crisis, and most of all, the many solutions that must be employed if we are to resolve the defining issue of our time. If we summon the human ingenuity and intelligence that have propelled us as a species, this can become a time of great opportunity. But we must act now. In an important first step, world leaders committed to tackle the problem in Paris in December 2015. If we follow through on those commitments and step up our efforts, we can create a healthier and more just world for everyone.

**DAVID SUZUKI**

Introduction

# BEYOND PARIS 2015

THE 2015 PARIS Agreement on climate change showed that the world is finally taking global warming seriously. Is it too little too late, or is there still hope for humanity? The agreement itself will only get us halfway to the emissions cuts experts say are necessary to avert devastating consequences, including rising sea levels, food and water shortages, and increased extreme weather, droughts, and flooding. But the fact that almost every nation, 195 plus the European Union, agreed to tackle the problem—and that most of them, including the two largest emitters, China and the U.S., have now formally ratified the Paris Agreement—offers hope, especially because the agreement calls on countries to regularly review and strengthen emissions-reduction targets. As important as talk and agreements are, it's time to put those words into action.

Despite worldwide commitments to address the crisis, climate change remains a contentious subject. Even the terms used to describe the phenomenon spark debate: Is it "climate change" or "global warming"? The scientific methods to determine the properties, impacts, and possible consequences of climate change are constantly scrutinized and discussed. Are the models accurate? Is it possible to draw definitive connections between individual extreme weather events and climate change? Do we understand all the various factors and the roles they play? How much does human activity contribute compared to natural processes? Will the consequences be catastrophic or could there be benefits?

There's also vigorous discussion around potential solutions. Can we find ways to burn fossil fuels cleanly? Should we deal with the causes or adapt? Do windmills kill too many birds? Is nuclear power the answer? And economic questions arise. Will cutting greenhouse gas emissions—such as carbon dioxide and methane, which create heat-trapping blankets around the planet—harm economies or create benefits? Is it even possible to address the problem under current economic systems? Many people are still confused about what the Intergovernmental Panel on Climate Change is or does. And, of course, there are those who deny that climate change even exists or insist that, if it does, it's not a problem, or it's a natural phenomenon over which we have no control.

All of these questions are good and necessary, but following the 2015 UN Climate Conference in Paris, we must accept that we have a problem that is in large part of our own making and that it will get worse without a collective global effort to address it. We must also keep in mind that it's not so much that the planet is in trouble; it's humanity. The planet and its natural systems are resilient; they recover. But the conditions that make the earth habitable and relatively comfortable for humans are jeopardized because of our actions.

Will getting the planet back on track be difficult? Yes, of course. But is it an impossible task? No.

Cars, air travel, space exploration, television, nuclear power, high-speed computers, telephones, organ transplants, prosthetic body parts... At various times these were all deemed impossible. I've been around long enough to have witnessed many technological feats that were once unimaginable. Even ten or twenty years ago, I would never have guessed people would carry supercomputers in their pockets—your smartphone is more powerful than all the computers NASA used to put astronauts on the moon in 1969 combined!

Despite a long history of the impossible becoming possible, often very quickly, the "can't be done" refrain is repeated over and over regarding global warming. Climate change deniers and fossil fuel industry apologists often argue that replacing oil, coal, and gas with clean energy is beyond our reach. The claim is both facile and false. Facile because the issue is complicated. It's not simply a matter of substituting one for the other. To begin, conservation and efficiency are key. We must find ways to reduce the amount of energy we use—not a huge challenge considering how much people waste, especially in the developed world. False because rapid advances in clean energy and grid technologies continue to get us closer to necessary reductions in our use of polluting fossil fuels.

It's ironic that anti-environmentalists and renewable energy opponents often accuse those of us seeking solutions of wanting to go back to the past, to living in caves, scrounging for roots and berries. They're the ones intent on continuing to burn stuff to keep warm—to the detriment of the natural world and all it provides. People have used basic forms of wind, hydro, and solar power for millennia. But recent rapid advances in generation, storage, and transmission technologies have led to a fast-developing industry that's outpacing fossil fuels in growth and job creation. Costs are coming down to the point where renewable energy is competitive with the heavily subsidized fossil fuel industry. According to the International Energy Agency, renewable energy for worldwide electricity generation grew from 13.2 percent in 2012 to 22 percent in 2013, with the share expected to increase to at least 26 percent by 2020.[1]

The problem is that much of the world still burns nonrenewable resources for electricity and fuels, causing pollution and climate change and, subsequently, more human health problems, extreme weather events, water shortages, and environmental devastation. In many cities in China, the air has become almost unbreathable. In California, a prolonged drought is affecting food production. Extreme weather events are costing billions of dollars worldwide.

We simply must do more to shift away from fossil fuels, and despite what the naysayers claim, we can. We can even get partway there under our current systems. Market forces often lead to innovation in clean energy development. But in addressing the very serious long-term problems we've created, we may have to challenge another "impossibility": changing our outmoded global economic system. As economist and Earth Institute director Jeffrey Sachs wrote in the *Guardian* in 2015, "At this advanced stage of environmental threats to the planet, and in an era of unprecedented inequality of income and power, it's no longer good enough to chase GDP. We need to keep our eye on three goals—prosperity, inclusion, and sustainability—not just on the money."[2]

Relying on free-market capitalism encourages hyperconsumption, planned obsolescence, wasteful production, and endless growth. Cutting pollution and greenhouse gas emissions requires conserving energy as well as developing new energy technologies. Along with reducing reliance on private automobiles and making buildings and homes more energy efficient, that also means making goods that last longer and producing fewer disposable or useless items so that less energy is consumed in production. And it means looking at agricultural practices and dietary habits, because industrial agriculture, especially meat production, is a major factor in global warming.

People have changed economic systems many times when they no longer suited shifting conditions or when they were found to be inhumane, as with slavery. And people continue to develop tools and technologies that were once thought impossible. Things are only impossible until they're not. We can't let those who are stuck in the past, unable to imagine a better future, hold us back from creating a safer, cleaner, and more just world.

This book can't provide all the answers, but it's an attempt to look at global warming from all angles: the science, the possible consequences, some of the barriers to resolving the problem, the political and economic implications, what is being done and what more could be done, and the potential solutions. In doing so, the aim is to contribute to a better understanding of the crisis and to offer hope for a better, cleaner future.

Although the book looks at the obstacles to resolving the climate question—including examining the work and motives of those who deny that climate change exists or that it is a problem or that we can do anything about it—it is written with acceptance of the research and knowledge of the vast majority of scientists and other experts worldwide who have determined over many decades (if not longer) of wide-ranging study and observation that global average temperatures are steadily increasing to dangerous levels and that human activity, mainly from fossil fuel combustion, agricultural practices, and destruction of forests and green spaces, is a major contributor.

As for the terms, although "climate change" and "global warming" are used interchangeably, some experts note a subtle difference: global warming is the overall phenomenon whereby global average temperatures are steadily increasing, whereas climate change is its result. That is, as global temperatures increase, climatic conditions change in various ways. Those changes can include increased precipitation in some areas and drought in others, extreme weather where it was once rare and milder weather where it was once more volatile—and even colder spells in some places.

Finally, someone will inevitably ask, "Why should we listen to what David Suzuki says about the subject? He's not a climate scientist." Why, indeed? It's true that my scientific background is not in climate science, but I am a scientist who understands the scientific method—how science is conducted, its uses, and its limitations. This book is also informed by massive amounts of research and writing by climate scientists and experts from around the world, including at the David Suzuki Foundation (DSF).

Of course, some of those who argue that only climate scientists can write and speak knowledgeably about global warming often won't accept any rational argument or scientific evidence for human-caused climate change and its consequences. To them, I can only repeat what former California governor and action movie star Arnold Schwarzenegger posted on Facebook just before the December 2015 UN climate talks: "Besides the fact that fossil fuels destroy our lungs, everyone agrees that eventually they will run out. What's your plan then?" He added, "A clean energy future is a wise investment, and anyone who tells you otherwise is either wrong, or lying. Either way, I wouldn't take their investment advice," noting that California at the time was getting 40 percent of its power from renewables and was 40 percent more energy efficient than the rest of the country, with an economy that was growing faster than the U.S. economy as a whole.

Schwarzenegger then offered a metaphor to sum up his argument:

> There are two doors. Behind Door Number One is a completely sealed room, with a regular, gasoline-fueled car. Behind Door Number Two is an identical, completely sealed room, with an electric car. Both engines are running full blast.
>
> I want you to pick a door to open, and enter the room and shut the door behind you. You have to stay in the room you choose for one hour. You cannot turn off the engine. You do not get a gas mask.
>
> I'm guessing you chose Door Number Two, with the electric car, right? Door Number One is a fatal choice— who would ever want to breathe those fumes?
>
> This is the choice the world is making right now.

To that I would add: What if the world listened to the naysayers and deniers and they turned out to be wrong? We would end up with rapid catastrophic consequences that threaten the very survival of the human species and many other life forms. What if we listen to all the world's leading scientists and experts who have studied the issue from every angle and it turns out they missed some key piece of information that overturns all of their theories? We will end up with cleaner energy; less pollution; probably stronger economies, as clean technology creates more jobs and wealth than resource extraction; and valuable fossil fuels left in reserve for a time when we learn to use them more wisely and less wastefully.

Finally, every effort has been made to ensure that the information in this book is accurate and up to date, but science and knowledge evolve, especially with regard to such a dynamic and complex subject that is being studied from so many angles by numerous experts. If nothing else, the hope is to provide a broad look at the current state of human-caused, or anthropogenic, climate change and the many solutions necessary to resolve it. With the Paris Agreement, the world committed to do something about global warming. Now it's time to act.

# PART 1

## The Crisis

SCIENTISTS HAVE KNOWN about the greenhouse effect and feedback loops for almost two hundred years, but it wasn't until the 1970s that the extent to which these phenomena were affecting global climate really started to sink in. By the mid-1980s, many scientists were warning that burning fossil fuels and pumping massive amounts of greenhouse gases such as carbon dioxide into the atmosphere were causing global average temperature to rise at a steady and alarming rate. In 1988, the World Meteorological Organization and United Nations Environment Programme established the Intergovernmental Panel on Climate Change to bring scientists from around the world together to examine the science and present results to government representatives. After the IPCC's *First Assessment Report* in 1990, and a supplemental report in 1992, the United Nations Framework Convention on Climate Change was established at the 1992 Earth Summit, or United Nations Conference on Environment and Development, in Rio de Janeiro, Brazil.

Subsequent international conferences were held and assessment reports were released in 1995, 2001, 2007, and

2014—each one providing stronger evidence for the human role in global warming and offering greater evidence of its potential disastrous consequences.

By the time world leaders signed the Paris Agreement in late 2015, based on the 2014 *Fifth Assessment Report*, there was no denying that humanity was facing a serious crisis, in large part of its own making.

As the scientific evidence for human-caused climate change has become more robust, our understanding of the consequences has grown. Using increasingly sophisticated computer models, tree ring and ice core samples, historical records, satellite data, observation, and other methods, scientists from a range of disciplines worldwide have examined global warming's current and potential impacts on everything from oceans to weather to agriculture to human health and beyond. It's not a pretty picture.

But just as our understanding of global warming and its consequences has grown, so too have the various solutions that we must employ to confront this great challenge. We know that we must cut down on and eventually stop burning fossil fuels for energy. That means shifting to cleaner, renewable energy sources such as solar, wind, and geothermal. We also know that we have to protect and restore green spaces, such as forests and wetlands, that absorb and sequester carbon and keep it from the atmosphere. And we have to look at agricultural practices and diets that contribute to climate change. The good news is that renewable energy technologies have been improving by leaps and bounds, with storage and grid systems making power from wind and sun increasingly viable. Awareness of how our own daily habits contribute to climate change and other environmental problems is also growing, leading many people to make changes that collectively add up to much-needed improvements.

If we understand the problem, have so much evidence for the catastrophic consequences that await us if we fail to address it, and have many solutions at hand, why are we so slow to address the crisis in a meaningful way? The 2015 Paris Agreement was a positive leap forward, but even it won't prevent the global temperatures from warming beyond the necessary threshold of 2 degrees Celsius over preindustrial levels, let alone the agreement's aspirational goal of 1.5 degrees Celsius. Many experts believe that even if all countries that signed the agreement meet their stated targets on time, global average temperatures will rise by 2.7 to 3.5 degrees Celsius—which scientists agree would be catastrophic for humans and numerous other species.

The complexity of the problem is a major barrier to resolving it. Because we have waited so long to address global warming, and because the $CO_2$ we've already emitted into the atmosphere will stay there for a long time, avoiding the worst impacts will require major changes in global governance, economic systems, and personal lifestyles. That needn't be bad. Addressing the challenge of climate change could have other benefits, such as reducing inequality, improving health and quality of life, sparking innovation, and creating jobs. But we need a paradigm shift in our economic thinking—which isn't as difficult as it might sound, if it's put into context. We've changed economic systems many times when they no longer met our collective needs or when they were seen to do more harm than good. But politicians and the people they are supposed to represent often resist change when it's easier to continue business as usual—especially when the success of politicians often depends on setting plans according to short-term election-cycle timelines rather than seeing the long-term big picture.

But complexity isn't the only barrier. The fossil fuel industry has become entrenched in all aspects of life globally, with massive infrastructure to support it. It's also the most profitable industry in history. It's no surprise, then, that people who profit from the exploitation of coal, oil, and gas would do what they can to protect their interests, even to the detriment of human health and life. Studies have shown that people in these companies have long known about climate change and its possible consequences but have done everything they can to downplay the problem, even spending large amounts of money and time on campaigns designed to sow doubt and confusion regarding the scientific evidence. Large power companies also see their interests threatened by distributed energy systems that can be employed at the household or community level.

The fact that wealthy industrialized nations have caused much of the problem while those that are starting to catch up want some of the same benefits also presents a barrier to progress. As countries with large populations such as China and India demand the same levels of comfort that fossil fuel use has conferred on wealthy industrialized nations, emissions increase. International trade deals that prioritize corporate profits over environmental protection and national interests also contribute to the worsening crisis.

The challenges and barriers are formidable but not insurmountable. Solutions are available, but often the political will to implement them doesn't match the urgency of the situation. With extremely powerful, wealthy, vocal opponents arguing that there is no problem, or that we needn't do anything about it if there is, many people are understandably confused, and progress is slowed or stalled. The following three chapters will examine what the evolving science and its history tells us about the increasing consequences of climate change, and explore many of the barriers to resolving the crisis.

# Chapter 1

# THE SCIENCE

CLIMATE SCIENCE HAS been around for a long time, and the physics behind phenomena such as natural feedback cycles and the greenhouse effect have been understood for close to two hundred years. The evidence that human activity—mainly burning fossil fuels but also agricultural and forestry practices—is contributing to rapid global warming that can't be explained entirely by natural causes has been building steadily over many decades, to the point of certainty today.

The problem is that many people don't understand the science; in fact, many don't even understand how science itself operates. Those who make massive profits or who benefit in other ways from maintaining the status quo often exploit this lack of understanding to convince people that climate change either isn't an issue or isn't one worth worrying about. This can be dangerous in an era when everyone with a computer has a public platform.

A common argument is that global warming is just a theory, not a fact—but this arises from a misunderstanding of scientific method. Science is based largely on hypotheses, theories, and laws. A hypothesis is an idea that has yet to be tested. A scientist may speculate on why something occurs or happens in a particular way. The scientist, or scientists, will then develop experiments and observations to test the hypothesis. If those experiments don't confirm the hypothesis, it's back to the drawing board. If they do, then the hypothesis could become a theory, or further experiments could be conducted to ensure that all factors have been taken into account.

A theory is based on a tested hypothesis or, more often than not, many hypotheses. Once experiments confirm that the hypotheses accurately describe and predict real-world occurrences, a theory is developed. Because science, understanding, and technology evolve, theories are often revised and occasionally, if rarely, disproven and discarded.

A scientific law describes a natural phenomenon and is often based on a mathematical formula. It doesn't explain how or why the phenomenon occurs. Like theories, laws can also be revised or overturned as new knowledge becomes available.

Because science is often about trying to disprove theories, our understanding of natural phenomena is constantly being tested. As the great physicist Albert Einstein pointed out, "No amount of experimentation can ever prove me right; a single experiment can prove me wrong."

This is especially true of a complex field like climate science. With so many variables, conditions, effects, hypotheses, and predictions, it is impossible to be 100 percent certain about any of it. But scientists are now about as certain as they ever get that the earth is warming at an unusually rapid pace and that humans are largely responsible. For the IPCC's *Fifth Assessment Report*, released in four chapters in 2013–14, hundreds of scientists and experts worldwide combed through the most up-to-date peer-reviewed scientific literature and other relevant materials to assess "the state of scientific, technical and socio-economic knowledge on climate change, its causes, potential impacts and response strategies."

They determined that it is "extremely likely," or 95 percent certain, that humans are a major factor in rapid global warming and that evidence for climate change itself is "unequivocal." Science rarely gets more certain than that, and the uncertainty only lies in the understanding that there may be undetermined factors or that natural factors could play a larger or smaller role than experiments and observation have illuminated. And, because a large part of climate science is predictive, there is room for variation. But all of the theories surrounding climate change have been and are being constantly tested, with scientists looking for flaws as well as ways that the theories can be confirmed. The overwhelming evidence shows that although the earth's climate constantly changes, it is now changing, warming, more rapidly than ever, and although natural phenomena such as solar and volcanic activity play a role in climatic changes, this rapid warming can only be explained by considering the major contribution of human activity. Increasingly sophisticated predictive models and observation also show that the extreme weather and other consequences we're experiencing now will only get worse if we continue to emit greenhouse gases into the atmosphere and damage or destroy the natural systems that absorb and store carbon.

As I'll show in the next section, this evidence has been building for much longer than many people realize.

## Ice Age Studies, Feedback Loops, and the Greenhouse Effect

SCIENTIFIC UNDERSTANDING OF the greenhouse effect isn't new. French mathematician and natural philosopher Joseph Fourier discovered in 1824 that the earth's atmosphere retains heat that would otherwise be emitted back into space by infrared radiation. Although he didn't call it the greenhouse effect, he explained his concept by comparing the earth and its atmosphere to a box with a glass cover. It's a simplistic comparison, and as the American Institute of Physics points out in an excellent summary of the history on its website (from which some of this section is drawn), a glass box or greenhouse does not function in entirely the same way as the earth and its atmosphere.[1]

Fourier's research inspired other scientists to consider the phenomenon. In 1859, Irish-English scientist John Tyndall began studying the ability of gases such as water vapour, carbon dioxide (then known as carbonic acid), ozone, and hydrocarbons to absorb and transmit radiant heat.[2] On finding that water vapour, ozone, and carbon dioxide, or $CO_2$, absorbed heat radiation better than gases such as oxygen, hydrogen, and nitrogen, he theorized that fluctuations in water vapour and carbon dioxide could affect global climate. He also discovered the idea of heat islands, by noting that the city of London was warmer than its surroundings.

Some years later, self-taught British scientist James Croll observed that dark surfaces such as soil, rock, and trees hold heat from the sun, whereas snow and ice remain cool, and that as a region cools, wind patterns change, which could affect ocean currents.

Much of the research to this time was aimed at under-standing the causes of ice ages. A major breakthrough in our understanding of the effect of greenhouse gases occurred in 1896. Croll's ideas led Swedish scientist Svante Arrhenius to surmise that a drop in Arctic temperatures could cause land that had been bare in summer to remain covered in ice year round.[3] This ice would reflect more of the sun's heat back into space, lowering the temperature even more, thus creat-ing a positive feedback cycle. He then observed that water vapour could also cause a feedback loop, as warmer air puts more water vapour into the atmosphere, and because water vapour holds heat in, more warm air is created. Because $CO_2$ also absorbs heat radiation, Arrhenius concluded that adding $CO_2$ to the atmosphere would contribute to this feedback cycle. Thus, burning fossil fuels and increasing $CO_2$ emissions into the atmosphere could increase water vapour, causing global average temperatures to rise.

Arrhenius wanted to understand what could cause an ice age, and his studies led him to conclude that cutting $CO_2$ in the atmosphere by half could cause one. But he also cal-culated what would happen if the amount was doubled by burning fossil fuels, and concluded that this would cause a 5-or-6-degree-Celsius increase in global average temperatures—an estimate surprisingly close to the one climate scientists came up with using much better computer models one hundred years later.

A year after Arrhenius published his findings, American geologist Thomas Chamberlin examined the earth's carbon cycles more deeply, and according to the American Institute of Physics, wrote that ice ages are "intimately associated with a long chain of other phenomena to which at first they appeared to have no relationship." It's a concept that indigenous peoples have taught me, and one that I often talk and write about: Everything is interconnected.

In his "very speculative" paper, published in 1897, Chamberlin hypothesized that $CO_2$ could affect feedback cycles that bring about ice ages. The complexity of his ideas involved looking at the effect of volcanoes as they spew $CO_2$ into the air, and what happens when volcanic activity is lower and carbon is absorbed and stored by minerals, plants, and oceans, called carbon sinks. Because the atmosphere contains only a small fraction of the earth's carbon compared to these carbon sinks, and carbon cycles through the atmosphere every few thousand years, Chamberlin proposed that climate conditions "congenial to life" are in a delicate balance.

At the time, however, it was believed that natural forces, such as solar activity and the ability of oceans to absorb and store carbon, were far more important factors and that $CO_2$ had an insignificant influence compared to water vapour. Many scientists believed climate was self-regulating and that small changes to atmospheric composition could not alter climate over brief time periods. Any $CO_2$ that human activity did emit into the air would be absorbed quickly by oceans (and, to some extent, forests and peat bogs)—which was mostly true at the time, when far smaller amounts of fossil fuels were being burned. Some also argued that excess atmospheric $CO_2$ would fertilize plants and create more lush life—which is also true, to a point. Although the notion that human activity, such as burning ever-increasing amounts of fossil fuels, could not affect a self-regulating climate has been thoroughly disproven by modern science, many people still make the same outdated arguments today.

As with earlier scientific investigations, most climate science in the first half of the twentieth century was driven by a desire to explain the causes of ice ages. In the 1950s, scientists started to get an idea of the bigger picture. In 1956, Maurice Ewing and William Donn, at New York's Lamont Geological Observatory, were also trying to explain ice ages, in particular the abrupt end of the most recent one. In looking at feedback cycles in the Arctic, they speculated that a complex set of circumstances could lead to rapid climate change over the next few hundred or thousand years. But the change they saw was the coming of another ice age.

Their theories were controversial and often criticized, but they did serve to spark a renewed interest in climate science, more testing of theories, and wider acceptance of the idea that changes in Arctic ice sheets and snow cover could cause rapid changes in planetary surface conditions.

By the 1950s, researchers in the Soviet Union were using this growing scientific knowledge to consider ways to deliberately alter local climatic conditions, by "making Siberia bloom by damming the Bering Straits, or by spreading soot across the Arctic snows to absorb sunlight," according to the American Institute of Physics. This led Leningrad climatologist Mikhail Budyko to examine the ways in which human influences could be amplified by feedback loops. As a result of his studies, he was one of the first scientists to raise concerns about the potential major effects of burning fossil fuels and other human activities. In 1961 and 1962, he published two books warning that growing energy use will warm the planet and cause the Arctic ice pack to quickly disappear, contributing to further feedback cycles.

In the mid-1960s, Budyko developed models that showed relatively small changes in global average temperatures and polar snow cover could cause feedbacks that would cause dramatic increases in temperature and sea levels. Researchers in Sweden, New Zealand, and the U.S. were arriving at similar conclusions. Although many of the studies pointed to a warming planet, some speculated that changes in solar activity or dust in the atmosphere could cause another ice age.

Over the next few decades, climate scientists developed increasingly sophisticated computer models to examine the effects of greenhouse gases on climate—especially as computer technology improved along with scientific knowledge. It also became easier to study other planets, such as Venus, which was covered in an atmospheric blanket of water vapour and $CO_2$, producing a massive greenhouse effect, and to examine past climatic events.

In 1973, a U.S. probe to Mars led the famous astronomer Carl Sagan and others to conclude that the red planet had undergone major shifts between cold and hot. Around the same time, analyses of seabed clay layers showed that Earth's ice ages had occurred in roughly 100,000-year cycles. Although these roughly matched calculations by Serbian scientist Milutin Milankovitch in the early twentieth century, research also demonstrated that Milankovitch's theories about the effects of subtle shifts in the earth's orbit were not sufficient to explain the massive changes. However, natural cycles including ice buildup and flow, warping of the earth's crust and sea level changes, combined with orbital shifts, could explain the ice age cycles.

Scientists also started looking into the effects of clouds, volcanic dust, smoke, and other aerosols on climate. Some initial

studies led researchers such as NASA's James Hansen to speculate that the world could be headed for a cooling phase. This
short-lived theory, which didn't take into account factors such
as ocean circulation, provides ammunition for climate change
deniers to this day.

By the late 1970s, many scientists were convinced that
Earth was getting warmer, but although many proposed
convincing hypotheses, no one was able to accurately and
definitively prove the cause. I spoke with science writer Isaac
Asimov about it in 1977 on CBC Radio's *Quirks & Quarks*.

By the 1980s, computers were becoming sophisticated
enough that it was possible to go beyond looking at climatic
conditions in isolation to examine the numerous intercon-
nections that can affect systems as a whole. More knowledge
was also being gleaned from seabed and ice cap core samples, which allowed scientists to examine regular ice sheet
advances and retreats over hundreds of thousands of years.
Although exact causes of current warming were still elusive,
many experts were starting to agree that the unusually rapid
warming they were seeing would bring about increases in the
incidence and severity of heat waves, flooding, droughts, and
storms—which did indeed start to occur worldwide. Because
warming was not uniform but rapid, causes such as solar activity could be ruled out.

By the late 1980s, the theory of global warming and its
human contributions had become well established in the scientific community. In 1989, the CBC television show I host, *The
Nature of Things*, did its first global warming program, and I also
hosted the five-part radio series *It's a Matter of Survival*, which
was in part about climate change. The response to the latter
(more than seventeen thousand letters in pre-email days) was

so overwhelming that my wife, Tara, and I decided we had to do more than just talk about environmental problems; we had to do something. So we gathered a group of people to discuss ideas and, out of that, the David Suzuki Foundation was formed in 1990.

As computer models and research methods improved, along with the body of scientific knowledge, complexity increased. How were biological systems affected by climate and $CO_2$, and how in turn did they affect climate and carbon? What impacts would all of this have on agriculture, forestry, and spread of disease?

By the 1990s, studies of the Arctic showed that twentieth-century warming was far greater and more rapid than anything seen in at least the past four hundred years.

Although scientific models and observations were by this time aligning, and most experts were able to confidently conclude that the planet was warming at an unusually fast rate, in part because of the greenhouse effect, the theories still had their critics. Convinced that natural self-regulation would overcome any human effects on climate over the long run, Massachusetts Institute of Technology (MIT) meteorologist Richard Lindzen set out to challenge the way models accounted for the effects of water vapour. Advances in satellite technology and data would later confirm the climate models and prove Lindzen wrong. He continued to look for flaws in the models, and although much of the modelling data were confirmed, his efforts at least made scientists work to improve models and to confirm data through other methods, including paleoclimate studies.

Numerous models with a wide range of varying parameters all confirmed that adding greenhouse gases to the atmosphere would cause global warming.

## Michael Mann and the Hockey Stick Graph

IN 1998, UNIVERSITY of Virginia climate scientist Michael Mann, with Raymond Bradley of the University of Massachusetts Amherst and Malcolm Hughes of the University of Arizona, examined paleoclimatic data by studying ice cores, tree rings, and corals, as well as more recent thermometer readings. In doing so, they were eventually able to reconstruct Northern Hemisphere temperatures going back one thousand years. Mann later worked with the University of East Anglia's Philip Jones to chart temperatures for the past two thousand years. They found conclusively that global mean temperatures spiked rapidly starting in the early twentieth century, just as industrial and other human activities were releasing ever-increasing amounts of $CO_2$ and other heat-trapping gases into the atmosphere.

The graph they created, which was used in the IPCC's *Third Assessment Report*, in 2001, looked like a hockey stick, with a long, steady line that took a sudden jump upward at the end. Although many other scientists confirmed the results, Mann's work became a target for those opposed to prevailing theories of anthropogenic climate change. Attackers included politicians, pundits, a few scientists, and two Canadians: former mining company executive and consultant Steve McIntyre and economist Ross McKitrick. Although subsequent research found that the Canadians were correct in pointing out some statistical errors, the failings were minor and did not significantly affect the overall results. Several studies found more serious errors in McKitrick and McIntyre's methodology, and dozens of subsequent studies using various methods and records have since confirmed Mann's original analysis, with only slight variations.

Another report by aerospace engineer Willie Soon and astronomer Sallie Baliunas, published in the journal *Climate Research*, claimed that the Northern Hemisphere was warmer during the medieval period than Mann estimated, but their methodology and data, and the publication's peer-review process, were found to be lacking, leading to the resignation of several of the journal's editors and an admission by the publisher that the article should not have been accepted as is. Soon has received much of his funding from fossil fuel companies, and Baliunas has been affiliated with a number of fossil fuel–funded organizations.

As Mann told *Scientific American* in 2005, "From an intellectual point of view, these contrarians are pathetic, because there's no scientific validity to their arguments whatsoever. But they're very skilled at deducing what sorts of disingenuous arguments and untruths are likely to be believable to the public that doesn't know better."[4]

## The IPCC and Global Efforts

IN RESPONSE TO the increasing knowledge—and alarm—about global warming, the World Meteorological Organization and UN Environment Programme set up the Intergovernmental Panel on Climate Change in 1988 at the request of member governments. According to the IPCC website, its goal was "to prepare a comprehensive review and recommendations with respect to the state of knowledge of the science of climate change; the social and economic impact of climate change, and possible response strategies and elements for inclusion in a possible future international convention on climate." Under its governing principles, its assessments were to be

"comprehensive, objective, open and transparent"; based on scientific evidence; and "neutral with respect to policy, although
they may need to deal objectively with scientific, technical and
socio-economic factors relevant to the application of particular
policies."

Its *First Assessment Report*, in 1990, provided much of the
impetus for the formation of the United Nations Framework
Convention on Climate Change (UNFCCC), "the key international treaty to reduce global warming and cope with the
consequences of climate change." It has since produced many
comprehensive assessment reports, including the 1995 *Second Assessment* that provided materials used by negotiators for
preparation and adoption of the Kyoto Protocol in 1997. The
*Third Assessment* was released in 2001 and the Fourth in 2007.
The IPCC was awarded the Nobel Peace Prize in 2007.

With each assessment, the science has become more robust,
and the number of scientists, writers, and contributors has
grown to include experts from around the world, with topics
covered becoming increasingly broad.

The *Fifth Assessment Report* was released from September
2013 to November 2014 in four chapters (1. current science,
2. impacts, 3. strategies to deal with the problem, and 4. a
final report synthesizing the three chapters). It showed more
scientific certainty than in 2007, when the *Fourth Assessment*
was released, that humans are largely responsible for global
warming—mainly by burning fossil fuels and cutting down
forests—and that it's getting worse and poses a serious threat
to humanity. It contained hints of optimism, though, and
showed that addressing the problem creates opportunities.

Scientists are cautious. That's the nature of science; information changes, and it's difficult to account for all interrelated

factors in any phenomenon, especially one as complicated as global climate. When they say something is "extremely likely" or 95 percent certain—as the *Fifth Assessment Report* did regarding human contributions to climate change—that's as close to certainty as science usually gets. Evidence for climate change itself is "unequivocal."

The first chapter alone cited 9,200 scientific reports in 2,200 pages, stating, "It is extremely likely that human activities caused more than half of the observed increase in global average surface temperature from 1951 to 2010." It also concluded that oceans have warmed, snow and ice have diminished, sea levels have risen, and extreme weather events have become more common.

The report also dismissed the notion, spread by climate change deniers, that global warming has stopped. It was thought to have been slowing slightly because of natural weather variations and other possible factors, including increases in volcanic ash, changes in solar cycles, and oceans absorbing more heat. But improvements in methods to measure sea surface temperatures led the U.S. National Oceanic and Atmospheric Administration (NOAA) to conclude in 2015 that oceans were warmer from 1998 to 2014 than previously thought and that a much-touted slowing or hiatus in warming didn't occur.[5] That study itself was challenged by a February 2016 study published in *Nature Climate Change*, which did find evidence of a slowdown in the rate of warming, though not a halt.[6] It also found the slowdown has probably ended. One thing the scientists and their studies confirm is that none of it means climate change is any less of a worry. In fact, the warmest ten years have all been since 1998 (itself an unusually warm year, and one that deniers have desperately cherry-picked as a starting point to claim that warming stalled), and in 2013, carbon dioxide levels rose by the highest amount in thirty years.

According to the IPCC *Fifth Assessment Report*, an increase in global average temperatures greater than 2 degrees Celsius above preindustrial levels would be catastrophic, resulting in further melting of glaciers and Arctic ice, continued rising sea levels, more frequent and extreme weather events, difficulties for global agriculture, and changes in plant and animal life, including extinctions. The report concluded we'll likely exceed that threshold this century, unless we choose to act. Subsequent research has shown that 2 degrees is too conservative and that warming over 1.5 degrees will probably lead to disaster. We're almost at 1 degree already!

The reasons to act go beyond averting the worst impacts of climate change. Fossil fuels are an incredibly valuable resource that can be used for making everything from medical supplies to computer keyboards. Wastefully burning them to propel solo drivers in cars and SUVs, and other inefficient energy uses, will ensure we run out sooner rather than later.

Nations working together to meet science-based targets to cut global warming pollution and create clean, renewable energy solutions would allow us to use our remaining fossil fuel reserves more wisely and create lasting jobs and economic opportunities. Energy conservation and clean fuels offer the greatest opportunities. Conserving energy makes precious nonrenewable resources last longer, reduces pollution and greenhouse gas emissions, saves consumers money, and offers many economic benefits.

Shifting to cleaner energy sources would also reduce pollution and the environmental damage that comes with extracting coal, oil, and gas. That would improve the health of people, communities, and ecosystems, and reduce both health care costs and dollars spent replacing services nature already provides with expensive infrastructure. Reducing meat consumption, which contributes to global warming, is also beneficial to human health.

The fast-growing clean energy and clean technology sectors offer many benefits. Improved performance and cost reductions make large-scale deployment for many clean energy technologies increasingly feasible. Worldwide spending on clean energy in 2013 was $207 billion.

By 2014, Germany, the world's fourth-largest economy, was getting a third of its energy from renewable sources and had reduced carbon emissions 23 percent from 1990 levels, creating 370,000 jobs.

## The 2015 Paris Agreement

FROM THE END of November to December 12, 2015, government ministers, negotiators, climate experts, and world leaders convened in Paris, France, to consider the implications of the IPCC's *Fifth Assessment Report* and to agree on how to deal with its findings. It may well have been the world's last chance for a meaningful agreement to shift from fossil fuels to renewable energy before ongoing damage to the world's climate becomes irreversible and devastating.

The nations that met in Paris are responsible for more than 95 percent of global emissions. Although it's far from perfect, the agreement they came up with marks a significant achievement. When nations last attempted a global climate pact—in 2009, at COP15 in Copenhagen, Denmark—negotiations broke down and the resulting declaration was considered a failure.

The Paris Agreement, in process and outcome, was a dramatic improvement—a product of the growing urgency to act on the defining issue of our time. It's the first universal accord to spell out ways to confront climate change, requiring industrialized nations to transition from fossil fuels to 100 percent renewable energy by 2050 and developing nations by about 2080.

Before meeting in Paris, governments drafted plans to reduce national carbon emissions beginning in 2020. One goal of the negotiations was to develop a review mechanism to encourage countries to improve targets over time. That was achieved, giving hope that reductions will keep global temperatures from rising more than 2 degrees Celsius above preindustrial levels. In fact, the newly revised limit of 1.5 degrees is acknowledged as a target for future goal setting. Although the commitments aren't enough to achieve either goal, improving targets every five years—as is called for in the pact—will get us closer. Past experience shows that once a commitment is made to address a crisis, many unexpected opportunities and solutions arise.

Still, getting the world back on track will not be an easy task, especially as it requires action on commitments from nations that haven't always lived up to their word. The world's largest greenhouse gas emitter, China, was criticized throughout the conference for trying to water down requirements for a common emissions-and-targets reporting system and opposing the requirement for countries to update emissions-reduction goals every five years, advocating instead for voluntary updates.

Compromises produced a final product that fell short of assigning liability for past emissions and providing dependable "loss and damage" payments to nations already suffering from the effects of climate change. And success can't be achieved without ongoing pressure to ensure targets are met and become more ambitious over time. Despite these shortcomings, the Paris Agreement was a leap forward in the fight against climate change. Funding for vulnerable and developing nations, plans to ratchet up ambition at regular intervals, and recognition of the role of indigenous knowledge will play major roles in future action.

The commitment should also inspire people at all levels of society to propose ways to speed up the shift to clean, renewable energy and reduce waste through greater energy efficiency. Although governments and industry must do a lot of the heavy lifting, it's up to all of us to ensure that the planet we want—with clean air, safe water, fertile soil, and a stable climate—stays within reach, for our sake and the sake of our descendants.

Of course, climate science goes beyond determining that the world is rapidly warming to catastrophic levels and that human activity is a major contributor. Scientific research and analysis are also examining the current and potential consequences of global warming, the impacts on various natural systems and human endeavours, the feedback loops that are created from those impacts, and the potential solutions, among other elements of this complex and disturbing phenomenon.

In the next chapter, we'll look at some of those impacts and what they mean for our world.

# Chapter 2

# CONSEQUENCES AND IMPACTS

**M**ASSIVE AMOUNTS OF scientific evidence, developed over many years by scientists around the world, confirm that human activity is contributing to unusually rapid global warming and that failing to address its causes could be catastrophic for humanity, as well as numerous other species that share this small blue planet. Many indigenous peoples have long known that everything in nature is interconnected, that what we do to one part of a natural system often has unforeseen consequences that cascade throughout the environment. Our survival depends on a complex interaction of air, water, land, and living things, all interconnected and interdependent. Yet we have been recklessly treating the planet and its atmosphere as little more than a source of raw materials and a dumping ground for our waste and emissions.

Part of the problem is that Western thought and science often view things in isolation. Because nature doesn't always behave the same in a lab, test tube, or computer program as it does in the real world, scientists and engineers have come up with many ideas that didn't turn out as expected.

The insecticide DDT was considered a panacea for a range of insect pest issues, from controlling disease to helping farmers. But we didn't understand bioaccumulation and biomagnification then—toxins concentrating up the food chain, risking the health and survival of animals from birds to humans. Chlorofluorocarbons, or CFCs, seemed so terrific we put them in everything from aerosol cans to refrigerators. Then we learned they damage the ozone layer, which protects us from harmful solar radiation.

The problems caused by DDT and CFCs were relatively easy to resolve, but we're now facing the most serious and complex unintended consequence ever: climate change, from destroying carbon sinks such as forests and wetlands, and from industrial agricultural practices—but mainly from burning fossil fuels.

Oil, gas, and coal are miraculous substances—energy absorbed from the sun by plants and animals hundreds of millions of years ago, retained after they died, and concentrated as the decaying life became buried deeper in the earth. Burning them to harness and release this energy opened up possibilities unimaginable to our ancestors. We could create machines and technologies to reduce toil, heat and light homes, build modern cities for growing populations, and provide accessible transport for greater mobility and freedom. And because the stuff seemed so plentiful and easy to obtain, we could build roads and vehicles—big cars that used lots of gas—so that enormous profits would fuel prosperous consumer-driven societies.

We knew fairly early that pollution affected human health, but that didn't seem insurmountable. We just needed to improve fuel efficiency and create better pollution-control standards. That reduced rather than eliminated the problem and only partly addressed an issue that appears to have caught us off guard: the limited availability of these fuels. But the trade-offs seemed worthwhile.

Then, for the past few decades, a catastrophic consequence of our profligate use of fossil fuels loomed. Burning them has released excessive amounts of carbon dioxide into the atmosphere, creating a thick heat-trapping blanket. Along with our destruction of natural carbon-storing environments, this has steadily increased global average temperatures, causing climate change.

We're now faced with ever-increasing extreme weather-related events and phenomena, such as ocean acidification, which affects myriad marine life, from shellfish to corals to plankton. The latter produce oxygen and are at the very foundation of the food chain. The cascading consequences of extracting and burning fossil fuels, as well as destroying carbon sinks, are affecting all facets of the planet—human health, climate, weather patterns and events, other species, oceans, agriculture, and a range of human activity. In turn, these consequences are interacting with other elements of nature and human existence, creating complex feedback loops and further unintended consequences.

The interactions are complex and numerous and we're still not clear on what some of the further consequences will be, but in this chapter, we'll examine some of the costs of upsetting the earth's carbon cycle.

## A World of Extreme Weather, Water, and Food

INCREASINGLY FREQUENT AND severe heat waves, heavy rain and snowfall events, hurricanes, tornadoes, wildfires, and floods—unusual weather and its effects are everywhere, and getting worse as the planet warms. California has been experiencing severe drought since 2011. Temperatures in Spain, Portugal, India, and Pakistan reached record levels in 2015, sparking wildfires and causing thousands of deaths and heat-related ailments. In the same year, heavy rains, flooding, and an unusually high number of tornadoes caused extensive damage and death in Texas, Oklahoma, and parts of Mexico.

The likely causes are complex: a stuck jet stream, El Niño, natural variation, and climate change. Even though it's difficult to link all events directly to global warming, climate scientists have warned for years that we can expect these kinds of extremes to continue and worsen as the world warms. Some hypothesize that the strange behaviours of 2015's jet stream and El Niño are related to climate change, with shrinking Arctic sea ice affecting the former.

Hurricane Sandy, which wreaked havoc on Caribbean nations and the U.S. East Coast in October 2012, offered a glimpse into our future.

Does that mean climate change caused Hurricane Sandy? Not necessarily. Experts know that tropical Atlantic storms are normal in fall. This one and its impacts were made unusually harsh by a number of converging factors: high tides, an Arctic weather system moving down from the north, and a high-pressure system off Canada's East Coast that held the storm in place.

But most climate experts are certain that the intensity of the storm and the massive damage it caused were in part related to changing global climate. Global warming causes sea levels and ocean temperatures to rise, which results in more rainfall and leads to a higher likelihood of flooding in low-lying areas. Scientists also believe 2015's record Arctic sea-ice melt may have contributed to the high-pressure system that prevented Sandy from moving out to sea. In short, the storm and the unprecedented flooding and damage are exactly what climate scientists have been telling us to expect as global temperatures rise.

Extreme weather events, including heat waves and drought, are no longer just model-based predictions. NASA scientist James Hansen, who sounded the alarm about climate change in 1988, wrote in the *Washington Post* in 2012, "Our analysis shows that it is no longer enough to say that global warming will increase the likelihood of extreme weather and to repeat the caveat that no individual weather event can be directly linked to climate change. To the contrary, our analysis shows that, for the extreme hot weather of the recent past, there is virtually no explanation other than climate change."[1]

A number of studies indicate a clear connection between increasing extreme weather and climate change.[2] One, by climatologists at the U.S. National Center for Atmospheric Research in Colorado, looked at rising global atmospheric and sea surface temperatures, which have increased water vapour in the atmosphere by about 5 percent since the 1950s. According to the 2015 paper, published in *Nature Climate Change*, "This has fuelled larger storms, and in the case of hurricanes and typhoons, ones that ride atop oceans that are 19 centimetres higher than they were in the early 1900s. That sea-level rise increases the height of waves and tidal surges as storms make landfall."

Because of the way the planet and its systems balance energy, extreme precipitation in some areas is increasing more quickly than overall rainfall. Rain is caused by water vapour cooling enough to condense into liquid when the atmosphere cools. But with more greenhouse gases warming the atmosphere, water vapour can build up, leading to heavy rain and snow events that are increasing in frequency and severity. With more moisture and energy in the atmosphere, and warmer oceans, the world can also expect more intense hurricanes. Besides flooding, extreme precipitation can cause crop damage and loss, soil erosion, water contamination, and more. Because water vapour is often held for longer by a warmer atmosphere, and released in extreme events over smaller areas, other areas often get less overall precipitation than before, leading to droughts in some places.

A Stanford University study found "accumulation of heat in the atmosphere can account for much of the increase in extreme high temperatures, as well as an average decrease in cold extremes, across parts of North America, Europe and Asia" but also concluded the influence of human activity on atmospheric circulation, another factor in climate change, is not well understood.[3]

What scientists do know is that moist air rises as it is heated near the equator. Once it is high enough to cool, the moisture falls as tropical rain. The dry air then moves north and south, normally dropping at the subtropics. With a warmer planet, it travels farther, causing drying conditions farther north and south. This could help explain ongoing drought conditions in California and other parts of the U.S. Southwest. A 2016 paper published in *Nature Climate Change* examined the effects of atmospheric circulation and increased water vapour on storms and flooding in England during winter 2013–14.[4] The researchers used model simulations and found that the historic precipitation and flooding were caused not just by increased moisture in the air but also by increases in the number of January days with westerly flow.

The damage that climate change is causing, which will get worse if we fail to act, goes beyond the hundreds of thousands of lives, homes, and businesses lost; ecosystems destroyed; species driven to extinction; infrastructure smashed; scarce or polluted food and water in many areas; and people inconvenienced. It will even devastate the one thing that many corporate and government leaders put above all else: that human creation we call the economy—the very excuse many of our leaders use to block environmental protection and climate action.

The U.S. Environmental Protection Agency (EPA) reported, "Between 2011 and 2013, the United States experienced 32 weather events that each caused at least one billion dollars in damages." According to Hansen, the Texas drought in 2011 alone caused $5 billion in damage. Repairing the damage from Hurricane Sandy in the U.S. is expected to cost at least $50 billion. And as former World Bank economist Nicholas Stern has pointed out, slowing climate change will cost us, but doing nothing will cost far more.

Earth is clearly experiencing more frequent extreme weather than in the past, and we can expect it to get worse as we burn more coal, oil, and gas, and pump more carbon dioxide and other greenhouse gases into the atmosphere. This can have profound and costly impacts on everything from agriculture to infrastructure, not to mention human health and life.

Increasing extreme weather threatens global water supplies and food security. We've already seen prolonged droughts affecting food supplies in what were once productive food-growing areas, such as California, parts of Africa, and elsewhere in the world. Flooding also destroys crops and degrades and erodes soil, and changing weather patterns—altered growing seasons, less predictable conditions, and shifting climate zones—are making food production a challenge. Melting glaciers and changes to the earth's hydrologic cycles will affect the availability of water for drinking and growing food. As the world has come to depend more and more on globalized food delivery, climate change will also put pressure on the ability to rationalize these systems. Transporting food over long distances will become increasingly difficult, unless we can find ways to ship products without burning large amounts of some of the most polluting fossil fuels. That means shifting to more local food production, but that is threatened by the difficulty many areas are experiencing, and will experience even more as the world continues to warm, to produce food for local populations. Increasing degradation and loss of productive soils because of industrial agriculture practices and climate change add to the problem.

As water and food security are compromised, disease can start to spread. Climate change isn't just a matter of more unpredictable weather or increased storms, precipitation, and heat waves; it threatens the basic elements we need to survive and be healthy.

## What Can We Learn from Arctic Ice?

THE ARCTIC MAY seem like a distant place, just as the most extreme consequences of our wasteful use of fossil fuels may appear to be in some distant future. Both are closer than most of us realize.

The Arctic is a focal point for some of the most profound impacts of climate change. As we saw in Chapter 1, much of our understanding of global warming comes from studies of the Arctic, where changing conditions can trigger feedback cycles that affect the entire planet. One of the world's top ice experts, Peter Wadhams of Cambridge University, calls the Arctic situation a "global disaster," suggesting ice is disappearing faster than predicted and could be entirely gone in the near future.[5] "The main cause is simply global warming: as the climate has warmed there has been less ice growth during the winter and more ice melt during the summer," he told the Guardian.[6]

Over the past thirty years, permanent Arctic sea ice has shrunk to half its previous area and thickness.[7] As it diminishes, global warming accelerates. This is caused by a number of factors, including release of the potent greenhouse gas methane trapped under nearby permafrost, and because ice reflects the sun's energy, whereas oceans absorb it.

Because of albedo feedback (which refers to the ability of a surface to reflect solar energy), ice-covered regions like the Arctic are affected to a greater degree than other areas by even small changes in global temperatures. Researchers say the Arctic is warming twice as fast as the rest of the planet. Fresh Arctic snow and ice can reflect as much as 80 percent of the sun's energy back to space, and melting ice in summer can reflect 50 percent. According to the U.S. National Oceanic and Atmospheric Administration, ocean water only has an albedo of 10 percent.[8] Even small amounts of warming cause ice and snow to melt, reducing the surface area that reflects solar energy. As more dark ocean and land surfaces are exposed, more energy is absorbed, which causes further warming, and further melting, and so on. These feedback loops in the Arctic are complicated, because the Arctic receives little or no sunlight during winter, but up to twenty-four hours of sunlight during summer. But warming during spring, summer, and fall causes spring melt to arrive earlier and fall freezing to start later, meaning the period during which solar radiation can be absorbed rather than reflected lasts longer. And because the oceans absorb more heat during summer, they release the heat during fall and into winter, causing the atmosphere to warm even more. Because the atmosphere over the Arctic is quite stable, the heat stays near the earth's surface, leading to amplification of warming in the area.

The increase in warming rates in the Arctic regions sets off another feedback loop, as $CO_2$ and methane, a greenhouse gas many times more potent than $CO_2$, are released from oceans, permafrost, and soils that are no longer frozen. This causes more warming and more melting, and so on.

According to the IPCC, Arctic warming and feedback loops will "contribute to major physical, ecological, sociological, and economic changes," including altered drainage, landscapes, species composition, marine ecosystems, and human communities.[9] Melting Arctic ice and subsequent warming will also cause sea levels to rise, and more rapid warming at lower latitudes as oceanic heat transfers are slowed.

With all we know about climate change and what's happening in the Arctic, you'd think world leaders would be marshalling resources to at least slow it down. Instead, industry and governments are eyeing new opportunities to mine Arctic fossil fuels. Factoring in threats to the numerous species of Arctic creatures—including fish, seabirds, marine mammals such as whales and seals, and polar bears—makes such an approach even more incomprehensible.

Royal Dutch Shell spent more than US$4.5 billion on operations and lease purchases in preparation for Arctic drilling.[10] But its record shows how risky this is. First, a spill containment dome failed a routine safety test and was crushed by underwater pressure. Later, a drilling rig, which was being towed to Seattle so that Shell could avoid paying some Alaskan taxes, broke free during a storm and ran aground on an island in the Gulf of Alaska. The disastrous BP oil spill in the Gulf of Mexico in 2010 showed how dangerous ocean drilling can be, even in relatively calm waters, and how bogus the claims of the industry are that it can contain or even clean up a spill.

Problems with exploration in the Arctic aren't new. In October 1970, a blowout at a natural gas well on King Christian Island in the Arctic Ocean created a massive flame as up to 5.7 million cubic metres of gas a day spewed for more than three months. It was the second blowout in the Arctic since drilling began the year before. Around the same time, the drilling consortium Panarctic Oils Ltd. was slapped with a huge fine for dumping junk steel, waste oil, and other garbage into the Arctic Ocean. The drilling companies found a novel solution to the latter problem: they convinced the Canadian government of the day to issue ocean-dumping permits, making the practice legal and common until 1993, when Inuit challenged one of the permits.

Of course, the worst danger is that increased exploitation of fossil fuel resources in the Arctic will exacerbate global warming. Responding to climate change and vanishing Arctic ice by gearing up to drill for the stuff at the root of the problem is insane. Unfortunately, many fossil fuel companies and governments are engaged in a mad rush to get as much oil and gas out of the ground—no matter how difficult—while there's still a market. The ever-increasing devastation of climate change means we will eventually have to leave much of the fossil fuels where they are—or at the very least, substantially slow the pace of extraction and use the resource more wisely—if we want to survive and be healthy as a species.

As Arctic ice melts, countries like Australia burn, and droughts, floods, and extreme weather increase throughout the world, it's past time to get serious about events in the Arctic and what they mean for global warming.

## Antarctica Tells Another Story

DOWN AT THE other pole, the effects of climate change are somewhat more complicated and less well understood. Warming is occurring at a slower pace than in the north, and some ice sheets appear to be shrinking while others may be growing. Geographical conditions explain some of the differences between global warming's effects on the two poles. The Arctic is an ocean surrounded by land, but Antarctica is a land mass surrounded by ocean. Because of that, sea ice is not as thick in Antarctica, and it moves more freely. Most of the sea ice that forms during Antarctica's winter melts in summer, whereas the Arctic retains more winter ice. Wind patterns and water currents also act differently between the two poles.[11]

A study in the *Journal of Glaciology*, led by Jay Zwally, chief cryospheric scientist at NASA's Goddard Space Flight Center, found that glacier mass in Antarctica's western region is declining while increased snowfall in the eastern interior has led to a "net gain of about 100 billion tons of ice per year," but other researchers have questioned those findings, which don't dispute global warming.[12]

"I don't think Zwally's estimates really matter so much in the grand scheme because adding a little snow to Antarctica in no way offsets the complete disintegration of the West Antarctic ice sheet in the near future," University of Alaska Fairbanks glaciology professor Erin Pettit said.[13]

As for the slower pace of warming in Antarctica, researchers from the University of Washington and the Massachusetts Institute of Technology say it's probably because gale-force westerly winds push surface water north, which pulls "deep, centuries-old water to the surface."[14]

And parts of Antarctica are warming rapidly. The U.S. National Snow and Ice Data Center explains that although most of Antarctica has yet to see dramatic warming, "the Antarctic Peninsula, which juts out into warmer waters north of Antarctica, has warmed 2.5 degrees Celsius (4.5 degrees Fahrenheit) since 1950."[15] However, parts of Antarctica are so cold that even if they heat by the same amount as the peninsula, it won't be enough to melt ice.

In any case, climate change is affecting Antarctica, and that has profound implications. The U.S. National Oceanic and Atmospheric Administration reported that atmospheric carbon dioxide reached four hundred parts per million in Antarctica on May 23, 2016—the first time it's passed that threshold in the remote area in 4 million years!

"The far southern hemisphere was the last place on earth where $CO_2$ had not yet reached this mark," said Pieter Tans, lead scientist of NOAA's Global Greenhouse Gas Reference Network. "Global $CO_2$ levels will not return to values below 400 ppm in our lifetimes, and almost certainly for much longer."[16]

Global warming's impacts on Antarctica are already negatively affecting some penguin populations and could have an impact on sea levels, as ice shelves collapse. Scientists are working to learn more about what is occurring in Antarctica and what the consequences might be, but the main lesson so far is that climate change knows no boundaries and impacts in Antarctica will be felt around the world.

## Oceans Take the Brunt of Global Warming

IT'S OFTEN SAID that we know as much about Mars and the moon as we do about oceans. Considering that oceans cover more than 70 percent of the earth, this should be cause for concern. At the very least, we should be doing more to protect oceans from the negative effects of human activities, including climate change, even if we don't fully understand all that is happening under the seas.

We do know, however, that greenhouse gas emissions have a tremendous impact on oceans. As thermal sciences professor John Abraham wrote, "As humans add more heat-trapping gases to the atmosphere, it causes the Earth to gain energy. Almost all of that energy ends up in the oceans. So, if you want to know how fast the Earth is warming, you have to measure how fast the oceans are heating up."[17]

Oceans and the life they support face numerous threats: pollution, overfishing, massive swirling islands of plastic waste, dead zones caused by nitrogen runoff from agricultural activities and sewage, acidification from excess $CO_2$, oxygen depletion, and more. No oceans have escaped the consequences of human activity. French scientists who completed a two-and-a-half-year journey covering more than eleven thousand kilometres through the Atlantic, Pacific, Antarctic, and Indian Oceans in 2012 found plastic debris in a remote ocean area that was thought to be pristine.

Researchers on the boat *Tara*, who were studying the effects of climate change on marine ecosystems and biodiversity, found plastic fragments in the Southern Ocean and Antarctica at levels comparable to the global average. "The fact that we found these plastics is a sign that the reach of human beings is truly planetary in scale," said Chris Bowler, scientific coordinator of Tara Oceans, in the *Guardian* in 2012.[18] It also reminds us that we live on a planet where everything is connected.

A 2011 study by the International Programme on the State of the Ocean (IPSO) found the combined effects of overfishing, fertilizer runoff, pollution, and ocean acidification from carbon dioxide emissions are putting much marine life at immediate risk of extinction.[19] The twenty-seven scientists from eighteen organizations in six countries who participated in the review of scientific research from around the world concluded that the looming extinctions are "unprecedented in human history" and have called for "urgent and unequivocal action to halt further declines in ocean health." The main factors are what they term the "deadly trio": climate change, ocean acidification, and lack of oxygen. Overfishing and pollution add to the problems.

Another study by the organization, in 2013, led IPSO scientific director Alex Rogers of Somerville College, Oxford, to conclude, "The health of the ocean is spiraling downwards far more rapidly than we had thought. We are seeing greater change, happening faster, and the effects are more imminent than previously anticipated. The situation should be of the gravest concern to everyone since everyone will be affected by changes in the ability of the ocean to support life on Earth."[20]

Ocean currents, upwellings, oxygen levels, acidity, and temperature are changing in ways we haven't seen before. Assumptions we once held about the seas are no longer valid. Oceans produce more than half the oxygen we breathe and absorb up to a third of carbon dioxide emissions, as well as providing an estimated annual economic value of at least $24 trillion.

Research compiled by the IPCC has described how ingredients in the ocean's broth are changing dramatically.[21] Life in the seas is closely linked to factors in the immediate surroundings, such as temperature; acidity, or pH; salinity; oxygen; and nutrient availability. These combine at microscopic levels to create conditions that favour one form of life over another and emerge into complex ecosystems.

Oceans now absorb one-quarter to one-third of the atmosphere's $CO_2$. That's good for the atmosphere but bad for organisms with calcium carbonate shells. While oceans help slow the pace of global warming, they too are absorbing too much carbon dioxide, resulting in disruption of the ocean's pH balance. This increasing acidity causes calcium carbonate to dissolve, affecting life forms including corals, shellfish, and several species of plankton that rely on calcium for their very structure. Organisms that form the base of the oceanic food chain, such as krill and shell-bearing zooplankton called pteropods, are at great risk, which puts all creatures higher up on the food chain, including humans, at risk. These organisms also store enormous amounts of carbon that will be released into the atmosphere as they die off. Even worse, phytoplankton produce much of the oxygen we breathe, and climate change is endangering these organisms. A 2015 study led by University of Leicester applied mathematics professor Sergei Petrovskii found that "an increase in the water temperature of the world's oceans of around six degrees Celsius—which some scientists predict could occur as soon as 2100—could stop oxygen production by phytoplankton by disrupting the process of photosynthesis."[22]

We're witnessing the effects of ocean acidification on shellfish along the West Coast of North America. In 2014, a Vancouver Island scallop farm closed after losing 10 million scallops, probably because of climate change and increasing acidity.[23] The U.S. National Oceanic and Atmospheric Administration has also linked oyster die-offs along the Pacific coast to climate change.[24]

As the IPSO points out, oceans play a key role in regulating the earth's climate and are subject to rising levels as global warming increases. Oceans absorb much of the heat caused by excess greenhouse gas emissions. In fact, better methods to measure sea surface temperatures led scientists at the NOAA to conclude in 2015 that oceans were warmer from 1998 to 2014 than previously thought and that a much-touted slowing or hiatus in warming didn't actually occur.[25] The 2013 IPSO study found that many negative changes to the oceans are occurring much faster than anticipated and continue to accelerate, either meeting or exceeding worst-case scenarios predicted by the IPCC and others. Arctic, Greenland, and Antarctic ice sheets are declining faster than expected, causing sea levels to rise more rapidly. This, in turn, is leading to "changes in the distribution and abundance of marine species; changes in primary production; changes in the distribution of harmful algal blooms; increases in health hazards in the oceans; loss of large, long-lived fish species causing the simplification and destabilisation of food webs in marine ecosystems," as well as increases in climate feedback loops.[26] The report concludes, "The longer the delay in reducing emissions the higher the annual reduction rate will have to be and the greater the financial cost. Delays will mean increased environmental damage with greater socio-economic impacts and costs of mitigation and adaptation measures."

The many other human-caused stressors on the oceans—including overfishing, pollution, agricultural runoff, and sewage—compromise the resilience of oceans in the face of climate change.

Oceans, land, and atmosphere are intricately connected to climate systems and changes. Wind and currents move warmer water toward the poles and cooler water toward the equator. Heat energy is transferred between the sun, atmosphere, land, and oceans through radiation, convection, and conduction. As levels of greenhouse gases such as $CO_2$, water vapour, methane, and ozone increase, warming occurs, with the oceans absorbing much of the warming. This affects ocean currents, and because warmer water expands and global warming causes glaciers and sea ice to melt, sea levels rise. Warmer oceans also alter climate patterns, increasing the frequency and severity of events like tropical storms. As the U.S. EPA points out, "Interactions between the oceans and atmosphere occur slowly over many months to years, and so does the movement of water within the oceans, including the mixing of deep and shallow waters. Thus, trends can persist for decades, centuries, or longer. For this reason, even if greenhouse gas emissions were stabilized tomorrow, it would take many more years—decades to centuries—for the oceans to adjust to changes in the atmosphere and the climate that have already occurred."[27] According to the EPA, global average sea levels have increased by about 1.5 centimetres per decade since 1880, but the rate has increased in recent years to 2.5 centimetres per decade. Sea level increases vary by region, and the EPA reports that increases have been as high as 20 centimetres between 1960 and 2014 along parts of the U.S. mid-Atlantic and Gulf coasts.

Even relatively small increases in sea level can cause shore-line erosion and changing coastal habitats, wetland destruction, contamination of agricultural areas and aquifers, and damage to human infrastructure. If climate change isn't curtailed, many highly populated areas will eventually be underwater. Because the oceans have already absorbed so much heat, scientists predict that sea level increases will displace 20 percent of the world's human population over the coming decades, even if global average temperature increases are kept below two degrees Celsius, from areas including Rio de Janeiro, New York City, Vancouver, London, Shanghai, and many others. Many low-lying islands will be completely submerged. Some studies predict sea levels could rise between 75 and 200 centimetres by 2100, or as much as 7 metres if the Greenland ice sheet were to melt.[28]

Increasing storm surges, rapid spread of invasive species and ocean-related diseases, and collapsing polar ice shelves are also consequences of warming oceans. And scientists believe that warming oceans could change global ocean currents that help regulate the world's temperature.

Although our knowledge of oceans and their role in global climate systems is continually improving, we still have a lot to learn. But we know enough to see that we have to start treating them differently if we are to survive and remain healthy. After all, we can't move to Mars or the moon.

## Climate Crisis Spells Trouble for Human Health

WHAT IF WE could reduce worldwide deaths from disease, starvation, and disaster while improving the health of people everywhere? According to the World Health Organization

(WHO), we can.[29] "Previously unrecognized health benefits could be realized from fast action to reduce climate change and its consequences," said a news release about the WHO's first global conference on health and climate in Geneva in summer 2014, adding, "changes in energy and transport policies could save millions of lives annually from diseases caused by high levels of air pollution." Encouraging people to use public transit, bicycle, and walk instead of driving would cut traffic injuries and vehicle emissions and promote better health through increased physical activity.

Studies show that heart attacks and respiratory illness because of heat waves, altered transmission of infectious diseases, and malnutrition from crop failures can all be linked to a warming planet. And economic and political upheaval brought on by climate change can damage public health infrastructure, making it difficult for people to cope with the inevitable rise in sickness. Research has also shown that warming ocean waters are increasing the incidence of waterborne illnesses, including those caused by toxic bacteria in shellfish.

Climate change affects the very basics that humans need to stay healthy and alive: clean air, safe water, productive and uncontaminated soils for growing food, and adequate shelter. According to the WHO, climate impacts on these basic needs will lead to 250,000 additional deaths a year, "from malnutrition, malaria, diarrhoea and heat stress" between 2030 and 2050. Weather-related natural disasters have more than tripled since the 1960s, resulting in more than 60,000 deaths a year, mostly in developing nations.[30] The young, elderly, and poor are at especially great risk.

This is costly to the economy as well as to human health and survival. The World Bank estimates that a severe influenza pandemic could cost the world economy $3 trillion. Environment Canada says air pollution alone costs the Canadian economy billions of dollars a year because of increased health care costs, missed workdays, and reduced productivity.

The 2015–16 spread of the Zika virus also gives us a glimpse of what to expect from climate change. Researchers believe the virus, which is transmitted by mosquitoes, could spread farther north as warmer, wetter weather provides ideal conditions for the mosquitoes to breed. The spread in South America is probably linked to wetter and warmer conditions there. Other mosquito-borne illnesses could also spread.

Reducing the threat of global warming and finding ways to adapt to unavoidable change will help people around the world "deal with the impact of heat, extreme weather, infectious disease and food insecurity," according to the WHO. Climate change affects human health in multiple ways. Increased extreme weather causes flooding and droughts, which influences food production, water, and sanitation. Pathogens that plague humans, livestock, and crops spread more widely. The WHO noted that diseases such as cholera, malaria, and dengue are especially sensitive to weather and climate changes: "Climate change is already causing tens of thousands of deaths every year from shifting patterns of disease, from extreme weather events, such as heat-waves and floods, and from the degradation of water supplies, sanitation, and impacts on agriculture." And it will get worse if we fail to address the problem.

Global warming and pollution also affect ailments such as asthma and allergies. A warming planet means longer growing seasons and stimulated plant growth in many areas (although it's causing drought and reduced plant growth in some parts of the world). Research shows the U.S. pollen season lengthened by about sixteen days from 1995 to 2014 and the ragweed season by anywhere from a day to sixteen days, with greater increases moving north. The Public Health Agency of Canada says Canada's ragweed season was close to a month longer in 2014 than in 1995 because of warming temperatures.

And rising atmospheric $CO_2$ actually increases pollen production. Tests conducted by U.S. Department of Agriculture weed ecologist Lewis Ziska showed pollen production doubled from five to ten grams per plant when $CO_2$ in the atmosphere went up from 280 parts per million in 1900 to 370 in 2000,[31] according to a USA Today article. That could double to twenty grams by 2075 if greenhouse gas emissions continue to rise.

Add to that the extreme weather impacts of climate change that can exacerbate allergy symptoms and other respiratory problems (rain and higher temperatures create more moulds and fungi in some places; more dust contributes to allergies and asthma in drought-stricken areas), plus the all-around increases in ground-level ozone, smoke, and pollution, and you've got a recipe for mass discomfort, illness, death, and rising health care costs.

We still don't fully understand the multiple impacts of global warming on allergies or what else may be contributing to the problem. Increased chemical exposure and the hygiene factor—which shows lack of exposure to germs and the outdoors early in life can make people more prone to allergies—may also be involved.

The effects on allergies don't mean people should stay indoors. Getting outside offers numerous physical and mental health benefits. Research even shows that kids who spend a lot of time outdoors develop fewer allergies. People can also take steps to minimize allergic reactions, such as going outside later in the day, when pollen levels are lower, and reducing allergens inside the home. After all, this isn't about plants being bad for people. We can't live without them. It's more about the natural systems that keep us alive and healthy being thrown out of whack by our reckless behaviour.

"The evidence is overwhelming: climate change endangers human health," said WHO director-general Margaret Chan. "Solutions exist and we need to act decisively to change this trajectory." Doing all we can to prevent climate change from getting worse will make life easier for all of us. If we want to protect our health, our children and grandchildren's health, and the natural systems that keep us alive and healthy, we must act.

## MENTAL HEALTH IN THE FACE OF CLIMATE DESPAIR

WHILE CLIMATE CHANGE is affecting and will continue to threaten human physical health, it's also taking a toll on mental health. People understandably feel afraid, grief-stricken, guilt-ridden, and often powerless to change what we are doing to the planet and its life-support systems. According to a 2016 Toronto Star article, "Signs of mental distress related to climate change have appeared in vulnerable populations, from drought-stricken prairie farmers to isolated aboriginal communities and the scientists who crunch climate data."[32]

The article points to a 2012 U.S. National Wildlife Federation report that concluded increasing heat waves, drought, extreme weather, and growing pressure on food and water systems and infrastructure will increase mental and social disorders, including "depressive and anxiety disorders, post-traumatic stress disorders, substance abuse, suicides and widespread outbreaks of violence," hitting "children, the poor, the elderly and those with existing mental health problems" the hardest.[33]

Climate scientists and environmentalists who study global warming are reporting increasing cases of depression, anxiety, stress, and other difficulties, coupled with the problems they bring about, such as marriage breakdown, substance abuse, and even suicide.

Although the article notes that the American Psychological Association is taking the issue seriously, studying and raising awareness about it, most health authorities are unprepared. Environmental lawyer David R. Boyd, who has done work with the David Suzuki Foundation, told the *Star* that he wrote *The Optimistic Environmentalist* in part to overcome the stress of examining environmental problems. "For me, writing this was a voyage of recovery," he said.[34]

## Birds and Other Animals Face Hard Times

OUR INSATIABLE ENERGY appetite puts many animals at risk. A 2015 report in *Science* magazine concluded that one in six animal and plant species could go extinct over the coming century if we don't do enough to address climate change, with those in South America, Australia, and New Zealand being hit particularly hard.[35] The researchers found that 2.8 percent of species are already at risk of extinction and that the risk would rise to 5.2 percent with a two-degree-Celsius increase in global temperatures. Extinction risks could be exacerbated by human activity, leading to habitat loss or damage, pollution, and alteration of natural systems by climate change. A 2013 study predicted that global warming could eliminate or deplete 82 percent of California's native freshwater fish species. Other studies have found that climate change will alter migration patterns and timing, affect reproduction, and make some fish smaller.

Birds face some of the highest risks. Reading some energy-related news and blogs, one might conclude wind power is the biggest bird killer. But that's far from true. Although poorly situated wind farms, especially ones using older turbine technology, do kill birds, it's an issue that can be addressed to a large extent with proper siting and good technology, as can problems around solar installations where birds have died. Fossil fuels, especially coal, are by far the largest energy-related bird killers. Heavy metals such as mercury and lead from burning coal kill numerous birds—and even change their songs, which can affect their ability to mate and protect territory.[36] And climate change is affecting many species' breeding and migratory patterns.

Renewable energy critics often cite the number of birds killed by wind power installations, but studies show that fossil fuel and nuclear energy are responsible for far more bird deaths—and house cats kill billions of birds a year.

Not only do birds fill us with awe and wonder, but they also provide food and feathers, and keep insects and rodents in check. Their ability to warn us of the drastic ways we're changing the world's ecosystems, climate, and water cycles can't be ignored. By working to ensure more species don't go the way of the passenger pigeon, we're also protecting ourselves from the effects of environmental destruction.

Habitat loss is a major threat for birds and other animals, and destroying green spaces where birds and animals live also reduces carbon sinks. According to many scientific studies, between sixteen thousand and seventeen thousand plant and animal species are threatened with extinction because of human activity, mostly through habitat loss. This includes 12 percent of all known birds, 23 percent of mammals, and 32 percent of amphibians. Climate change is predicted to sharply increase the risk of species extinction within our own children's lifetime. According to the IPCC, 20 to 30 percent of plant and animal species assessed will probably be at increased risk of extinction if global average temperatures continue to rise with escalating emissions of carbon pollution.

Animals are especially at risk in the Arctic, where global warming is occurring more rapidly than elsewhere, with more severe consequences. The international community has flagged global warming as a major threat to the survival of polar bears. In 2006, the International Union for Conservation of Nature, or IUCN, listed the polar bear as a "vulnerable" species. In a 2015 update of its Red List of Threatened Species, the IUCN flagged "loss of sea ice habitat due to climate warming as the single most important threat to the long-term survival of the species."[37] Polar bears, which are extremely important to the culture and livelihoods of indigenous people, travel over sea ice to get to their prey. The IUCN update says, "An annual ice-free period of five months or more will cause extended fasting for the species, which is likely to lead to increased reproductive failure and starvation in some areas. According to recent sea ice projections, large regions of the Canadian Arctic Archipelago will be ice-free for more than five months by the late 21st century; and in other parts of the Arctic, the five-month ice-free threshold may be reached by the middle of the 21st century. Warming Arctic temperatures could also reduce habitat and increase the incidence of disease for prey species such as ice seals, placing the polar bear at further risk."[38] The bears also face threats from pollution, resource extraction, oil spills, and other development. As top predators, polar bears help keep northern ecosystems in balance. The IUCN is working with Canada, Norway, Greenland, Russia, and the United States on a Circumpolar Action Plan to help the bears survive.

The IUCN says 35 percent of bird species, 52 percent of amphibians, and 71 percent of reef-building corals are "particularly vulnerable to the effects of climate change."[39] Scientists like famed Harvard entomologist E.O. Wilson have described the current wildlife crisis as a silent epidemic, because it receives so little attention from governments. As individuals, we must conserve energy, shift to cleaner sources, and demand that our industrial and political leaders address issues such as pollution and climate change. And we can work to protect wetlands and other bird habitat. We can also join the legions of citizen scientists who are contributing to avian knowledge by posting information to sites such as eBird.org.

## Climate Change Exacerbates Conflict and Refugee Crises

CONFLICTS AND REFUGEE crises aren't new, but 2015 was marked by particularly devastating events. As world leaders and experts prepared to meet in Paris to address the climate crisis, refugees from the horrific conflict in Syria were fleeing for camps in neighbouring countries and into Europe, many of them not making it in their shoddy and overcrowded boats.

Just as it's not always possible to definitively connect one extreme weather event entirely to climate change, it would be wrong to blame the Syrian crisis solely on climate change. But many analysts have noted a connection. Although the country has been beset by conflict over differing political and religious ideologies, as well as resources, many experts note that drought and increasing water scarcity—caused in large part by climate change—forced many people to flee from agricultural areas to cities. Along with an influx of Iraqi war refugees, that caused Syria's urban population to increase from "8.9 million in 2002, just before the U.S. invasion of Iraq, to 13.8 million in 2010," according to an article in *Scientific American*.[40] The article quotes a 2015 study in the *Proceedings of the National Academy of Sciences of the USA*: "The rapidly growing urban peripheries of Syria, marked by illegal settlements, overcrowding, poor infrastructure, unemployment and crime, were neglected by the Assad government and became the heart of the developing unrest."[41]

Poor management and political decisions before and during the crisis have exacerbated the problem, but the climate connection is being seen in many conflict-ridden areas.

As some parts of the world heat up and experience increased drought or flooding, with subsequent damage or devastation to agricultural systems, more refugees will leave increasingly uninhabitable areas, or areas where they can no longer make a living or grow food. As in Syria after the 2007–10 drought, many will make their way to already overburdened cities. Still more will flee to countries where the effects of climate change aren't as bad or where infrastructure makes it easier to cope with the consequences.

Organizations including the International Red Cross, World Bank, and United Nations High Commissioner for Refugees report that "environmental refugees" now outnumber political refugees and that the problem will get worse as the effects of climate change create harsh or even unlivable conditions in many parts of the world. And, unlike political refugees, environmental refugees are not protected by international law. Many are fleeing from rural or coastal areas to urban centres in their own countries.[42]

Scientists say that drought and desertification affect almost 30 percent of Earth's land surface and threaten the well-being of more than a billion people worldwide. Although the cumulative effects of overgrazing, over-cultivation, deforestation, and poor irrigation are factors in desertification, so too are climate change and extreme weather. The deterioration of dry-land ecosystems has already created desert-like dead zones that can no longer support human life in places such as Sub-Saharan Africa. No region is immune. Close to three-quarters of North America's dry lands, including parts of the prairies, are vulnerable to drought. Many people have already been displaced from areas around China's expanding Gobi Desert, northwest Africa's Sahara Desert, and the Horn of Africa, among other places.

Sea-level rise is also displacing a steadily increasing number of people. In Bangladesh, half a million people were left homeless when rising sea levels started to submerge Bhola Island. According to National Geographic, "Scientists predict Bangladesh will lose 17 percent of its land by 2050 due to flooding caused by climate change. The loss of land could lead to as many as 20 million climate refugees from Bangladesh."[43] Other coastal areas and island nations, including many places in North America and Europe, will also be affected.

Throw political conflict into the mix and the situation becomes even more volatile. Studies by the UN and others have concluded that drought and environmental degradation from climate change, which caused rapid spikes in food prices, probably contributed to the 2010 Arab Spring uprising and the 2007 Darfur conflict. As fossil fuels become increasingly scarce and difficult to obtain, conflict will also be likely to increase in already volatile areas where those resources are located.

All of this is occurring as the world has only reached warming of less than 1 degree Celsius above preindustrial levels. Failing to limit global average temperature increases to 2 degrees, or the 1.5 degrees called for in the Paris Agreement, could have absolutely devastating consequences. According to the World Bank, an increase of 2 degrees would increase extreme heat days from four to sixty-two days in Amman, Jordan; from eight to ninety days in Baghdad, Iraq; and from one to seventy-one days in Damascus, Syria.[44] In Beirut, Lebanon, and Riyadh, Saudi Arabia, "the numbers of hot days are projected to reach 126 and 132 days per year respectively." Those numbers will rise significantly if world temperatures go above 2 degrees. The increased number of hot days, along with decreasing rainfall, will have serious impacts on water availability and agriculture, with corresponding spikes in food prices. This can only increase conflict and the number of people fleeing for more hospitable territories.

## Could Hockey Become an Endangered Sport?

WITH EVER MORE frequent droughts and floods causing food and water problems, rising sea levels pushing up property losses and infrastructure costs, and extreme weather and pollution increasing illness and death, hockey may be the least of our worries when it comes to climate change. But outdoor winter sports are important economically and culturally, and provide ways for people to stay active and healthy during winter.

Unfortunately, hockey could well be a casualty. Research from Montreal's McGill and Concordia Universities shows global warming is having an effect on outdoor rinks in Canada.[45] "Many locations across the country have seen significant decreases in the length of the oss [outdoor skating season], as measured by the number of cold winter days conducive to the creation of rink ice," states their 2012 study. "This is particularly true across the Prairies, and in Southwest Canada, which showed the largest (and most statistically significant) decreases in the calculated oss length between 1951 and 2005."

This echoes a 2009 David Suzuki Foundation report, *On Thin Ice: Winter Sports and Climate Change.*[46] The McGill investigation looked at constructed outdoor rinks, whereas the DSF's focused on frozen rivers, canals, and lakes, but the conclusions are similar. Both predict that, unless we rein in greenhouse gas emissions, outdoor skating in parts of Canada could be history within the next fifty to one hundred years (the McGill study's authors now say it could happen within twenty to thirty years), and the length of the outdoor skating season will continue to shorten across the country.

Meanwhile, at Ontario's Wilfrid Laurier University, geographers have launched www.RinkWatch.org, a website where people can record information about backyard or neighbourhood rink conditions over the winter. "Our hope is that Canadians from coast to coast will help us track changes in skating conditions, not just this year, but for many years to come," associate professor Robert McLeman said in a release. "This data will help us determine the impact of climate change on winter in terms of length of season and average temperatures."[47]

According to the DSF report, one of Canada's best-loved outdoor skating venues, Ottawa's Rideau Canal, provides an example of what to expect. The report concludes that, with current emissions trends, the canal's skating season could shrink from the previous average of 9 weeks to 6.5 weeks by 2020, less than 6 weeks by 2050, and just 1 week by the end of the century.

On Thin Ice noted that many of Canada's hockey heroes got their start on outdoor rinks. "Without pond hockey, we probably wouldn't have what has become the modern game of hockey," the authors state. The DSF study says climate change could have a profound effect on many other winter sports, from skiing and snowboarding to winter mountaineering. But losing winter recreation opportunities, let alone our ability to produce food and keep our homes warm and people healthy, needn't happen. Taking action to avoid the worst impacts of climate change will ensure that kids and adults alike can continue to skate and score goals and enjoy winter in so many other ways.

# Chapter 3

# OBSTACLES AND BARRIERS

**A**s WE SAW with the 2015 UN Climate Conference and subsequent Paris Agreement, experts and world leaders are taking the issue of global warming seriously. Although some may quibble over details around exact causes, effects, and implications, very few legitimate scientists and researchers deny that human activity is a major contributor to global warming. No peer-reviewed scientific study has overturned the prevailing theory of climate change. Scientific research has also found conclusive evidence for the causes, which include burning fossil fuels, damaging or destroying forests and other green spaces, and employing energy-intensive and methane-producing agricultural practices that contribute to rising emissions. Although the science of global warming and all its related phenomena—from feedback cycles to impacts on oceans, land, ice, and snow; animal migration patterns; and more—is incredibly complex, the major causes are not, and so the solutions are not as complicated as they seem. We know that to address the problem we must conserve energy and find cleaner ways of obtaining energy than burning fossil fuels, and we must protect and preserve the natural systems that absorb and store carbon, such as oceans and forests. And although excessive consumption in the developed world contributes as much or more to climate change than population growth in the developing world, stabilizing population growth must also be part of the solution. The best way to achieve that is by giving women greater rights worldwide, especially regarding access to birth control and family planning education, as well as the right to participate in society and political decision-making.

If we agree there's a problem and we know what to do about it, why are we still pursuing activities that contribute to it? Why aren't people around the world doing more to slow or halt global warming? The 2015 Paris Agreement was a start, but even it doesn't put us on track to achieve the level of emissions reductions needed to avert disaster. A big part of the problem is that, for many reasons, we've failed to do enough for so long that it has become more difficult than ever to resolve what has become a global crisis. We've developed massive infrastructure for a fossil-fuelled world, especially in industrialized nations. Fossil fuels have made possible much of our prosperity, technological advances, breakthroughs in agricultural methods and scientific developments, and increasingly convenient lifestyles. It's not that easy to retool or quickly shift direction when fossil fuels have become such an integral part of human existence. Fossil fuels have also, in part, driven the world's population boom. And as poorer nations with large and often rapidly growing populations come to expect the same benefits people in the industrialized world have enjoyed, they look to inexpensive and relatively easily obtained fossil fuels to help them catch up.

Politicians and the systems they operate within are geared more toward short-term than long-term planning. Immediate economic prosperity keeps people happier and, in democracies, wins more votes than long-term planning that might include sacrifices. And a country like China, the world's largest greenhouse gas emitter, has a political system that allows it to say one thing and do another, with few ways for its people or the rest of the world to know what it's actually doing in terms of increasing or reducing its contribution to climate change.

The fossil fuel industry, which is by far the most profitable industry in the history of humanity, has also spent enormous amounts of time and money to protect its interests. That has involved political lobbying, paying massive amounts of cash to politicians and political parties to further its interests, campaigning to sow doubt and confusion about global warming and its consequences—often through front groups with scientific-sounding names or bogus studies—and developing infrastructure that locks societies into fossil fuel use.

This chapter will examine some of the barriers to resolving the climate crisis, including many of the arguments used to downplay the problem and its severity.

## The China Syndrome

THOSE WHO REJECT the evidence for anthropogenic climate change throw many slippery arguments onto the path of progress. Most of them are irrational, easily debunked, or disingenuous, but one contains a germ of truth. "Unless China and India stop emitting such high levels of greenhouse gases, nothing other countries do will make a difference," they claim. It's used as an argument against taking action on climate change, the assumption being that China and India will continue to increase emissions as their populations grow and industrialization expands.

Every nation, no matter how great or small its contribution to global warming, must do its part to resolve the problem, regardless of what other nations may or may not do. Arguing that we should give up because others aren't doing their part is an immature and facile argument. But it's true that if China and India refuse to do their part, we're pretty much screwed.

China is the largest greenhouse gas emitter, accounting for one-fifth or more of the world's energy consumption and close to one-quarter of emissions—and its energy use is steadily increasing. What it does or doesn't do to combat climate change will make a big difference in global efforts. (Although India is a large and heavily populated country with the world's fastest-growing economy, it is the fourth-largest greenhouse gas emitter, behind China, the U.S., and the European Union, in part because its per capita emissions are much lower than in the U.S., China, the EU, and many other countries.)

With Beijing's 21 million residents living in a toxic fog of particulate matter, ozone, sulphur dioxide, mercury, cadmium, lead, and other contaminants, mainly caused by factories and coal burning, you'd think the government would have many incentives to address the problem. Schools and workplaces regularly shut down when pollution exceeds hazardous levels. People have exchanged paper and cotton masks for more elaborate filtered respirators. Cancer has become the leading cause of death in the city and throughout the country. But it is possible to tackle some of the worst and most visible elements of pollution, such as particulate matter, without having much of an effect on overall greenhouse gas emissions—though pollution and climate change are intricately connected, despite what the Chinese government and some Western anti–climate science organizations have disingenuously claimed. Still, scrubbers and filters on smokestacks, as well as methods to capture some of the pollution, will decrease particulate matter, which is nasty and visible. But it's impossible to tackle the bulk of the pollution problem without addressing its primary causes—and in China, that's mainly burning coal for power generation and industrial uses, as well as oil for transportation. These are also major factors in climate change.

Chinese authorities, often reluctant to admit the extent of any problem, can no longer deny the catastrophic consequences of rampant industrial activity and inadequate regulations. According to *Bloomberg News*, Beijing's Center for Disease Control and Prevention says that, although life expectancy doubled from 1949 to 2011, "the average 18-year-old Beijinger today should prepare to spend as much as 40 percent of those remaining, long years in less than full health, suffering from cancer, cardiovascular disease, and arthritis, among other ailments."[1]

China's government also estimates that air pollution prematurely kills from 350,000 to 500,000 residents every year. Water and soil pollution are also severe throughout China. The documentary film *Under the Dome*, by Chinese journalist Chai Jing, shows the extent of the air problem. More than 150 million Chinese viewed the film in its first few days, apparently with government approval. Later it was censored, showing how conflicted authorities are over the problem and its possible solutions. The pollution problem also demonstrates the ongoing global conflict between economic priorities and human and environmental health.

Although China was criticized during the 2015 Paris climate negotiations for trying to water down requirements for a common emissions-and-targets reporting system and opposing a process to require countries to update emissions-reduction goals every five years, advocating instead for voluntary updates, it did shift to a much more constructive position in Paris than it had taken at previous climate negotiations—and has since formally ratified the Paris Agreement. China's government promised to cap $CO_2$ emissions and shift to about 20 percent alternative fuels (including nuclear) for energy by about 2030, reduce carbon and energy intensity, and cap coal consumption in some provinces. It has also promised to install an environmental tax and a cap-and-trade system for $CO_2$ in 2017. (Cap-and-trade systems set an overall limit, or cap, on the amount of greenhouse gas emissions a jurisdiction [province, state, or country] can emit. Polluters, such as heavy industry and electricity generators, are given permits or "allowances" that determine the level of emissions they can produce. Companies can buy, or trade for, allowances from more efficient firms that have more allowances than they need.)

Some observers are optimistic, going as far as to say that China under-promised so that it can over-deliver. Others are more skeptical, noting that Chinese authorities are notorious for saying one thing and doing another, and that the country's closed political system doesn't lend itself to independent monitoring or enforcement. China also approved 155 new coal-fired power plants in 2015, a 55 percent increase over 2014 and more than the number of approvals in the previous three years. China is also a major exporter of coal-fired power plants. And, as China watcher Elizabeth Economy wrote in

*Forbes* in 2015 regarding cap and trade, "China's non-market economy and weak rule of law pose a real challenge to such a system, given that cap-and-trade is designed to operate within a transparent, rules-based market system."[2] Economy added that China lacks "an up-to-date inventory on emissions that will allow Beijing to set the original cap of allowances properly; transparency in the auctioning process as well as in the amount of emissions; a robust system for monitoring emissions; a law that governs rules for a cap-and-trade system... and a penalty structure for non-compliance that is tough enough to induce behavioral change."

As organizations like the International Renewable Energy Agency, or IRENA, have also pointed out, China's reporting on energy, including coal and renewable energy, is not always trustworthy, especially when it comes to capacity versus actual deployment. But the IRENA notes that China is already a global leader in renewable energy development and has the capacity to significantly step up its game, thanks in part to "increasing cost-competitiveness of renewable energy technologies and other benefits such as improved energy security and decreased air pollution."[3]

Although burning coal is much cheaper than employing renewable energy sources, the IRENA notes that when factors such as pollution and health impacts are added, renewables start becoming more attractive. For example, reaching a 26 percent share for renewable energy by 2030 (as opposed to the business-as-usual estimate of 16 percent) would require investments of US$145 billion a year between 2014 and 2030, but savings from reduced $CO_2$ emissions and improved public health would range from US$55 billion to US$228 billion. The IRENA also notes that China will have to make significant improvements to its grid and transmission infrastructure to take advantage of increased renewable energy.

China is often accused of looking after its own interests at the expense of global interests, but when it comes to energy and climate, the IRENA makes a compelling argument that China has a lot to gain from shifting its power system: "If China acts decisively to increase the role of renewables in its energy system, it can significantly reduce the pollution of its environment, enhance its energy security, benefit its economy and play a leading role in mitigating climate change."

Rather than seeing China's situation as a warning, many people in Canada and the U.S.—including in government— refuse to believe we could end up in a similar situation here. And so U.S. politicians fight to block pollution-control regulations and even to remove the power of the Environmental Protection Agency, or shut it down altogether! In Canada, politicians and pundits have argued that environmental protection is too costly and that the economy takes precedence.

As noted earlier, some people even point to China as a reason for less populated countries like Canada not to do anything, arguing that what Canada and other smaller countries do or don't do to confront climate change and pollution will make little difference because their contributions pale in comparison to countries like China and India. But although Canada's air quality is better than many places, half of all Canadians live in areas where they are exposed to unsafe levels of air pollution. According to the Heart and Stroke Foundation, "Short and long term exposure to air pollution are estimated to result in 21,000 premature deaths in Canada in 2008 as well as 620,000 doctor visits, 92,000 emergency department visits, 11,000 hospital admissions and an annual economic impact of over $8 billion."[4] Canadians are also among the highest per capita greenhouse gas emitters.

And, as we know, air doesn't stay within national boundaries. The global atmosphere is being loaded with the sum of all nations' activities.

## Political Intransigence

WITHOUT GENUINE AND verifiable action on its climate commitments, and without steadily improving emissions-reduction goals and other mitigation strategies, China could be the biggest hurdle to avoiding the worst consequences of climate change. Many people are rightfully skeptical about the country's will to pull its weight, in light of its authoritarian government, lack of transparency, and reputation for protecting the interests of the ruling party elite over all else.

But transparency and democratic traditions are no guarantee that a country will fulfill and improve its climate commitments. We just have to look to the world's second-largest greenhouse gas emitter, the United States, to understand this. The U.S. was estimated to be responsible for about 16 percent of global $CO_2$ emissions from fossil fuels and some industrial processes in 2011, according to the Environmental Protection Agency.[5] Of the top ten countries for greenhouse gas emissions, the U.S. is second only to Canada for per capita emissions.[6]

Although commitments and programs under the Obama administration showed the U.S. government to be taking climate change seriously, opposition to progress has been confounding. Every leading Republican contender for the 2016 presidential election was on record as rejecting the evidence for climate change or the need to address it. According to *Mother Jones* magazine in 2015, "Ted Cruz believes that climate change is a 'pseudoscientific theory'; Donald Trump calls it a 'hoax'; and Ben Carson insists there's 'no overwhelming science' that it's caused by humans."[7] Cruz said, if he were elected, he would pull the U.S. out of the Paris Agreement. It's astounding that those who want to lead the most powerful country in the free world would display such ignorance and disregard for the future of the people they hope to represent and for all of humanity. It can't be explained entirely by the massive donations they receive from fossil fuel interests—though that appears to be a factor, which is disturbing.

With the Republicans under Donald Trump winning the November 2016 presidential election, the challenges will be difficult and numerous.

Elections can quickly change a country's approach to issues like climate change. Canada had been doing little at the federal level to address the problem, and was criticized for close to a decade for obstructing progress at international talks. Within weeks of winning the October 19, 2015, election, the new Liberal-led government had put together a high-profile cabinet committee on environment, climate change, and energy; expanded the minister of environment's title to include climate change; and appointed a minister of science and a minister of innovation, science, and economic development. The government invited opposition party and provincial and territorial leaders to the Paris talks, and was seen as taking a constructive approach to negotiations. At the same time, Canada's leaders still talk about the need for more pipelines to get Alberta's bitumen to market and have approved fossil fuel infrastructure projects, including a liquefied natural gas terminal near Vancouver. It's as if they don't really understand the seriousness of climate change or its causes.

The free-market fundamentalism that now informs much of the world's thinking and behaviour on economics and trade is such a significant hurdle that many of its adherents and proponents refuse to even accept that climate change is a problem worth addressing. Doing so would mean admitting failure. As Naomi Klein writes in *This Changes Everything*, "The traditional political left does not hold all the answers to this crisis. But there can be no question that the contemporary political right, and the triumphant ideology it represents, is a formidable barrier to progress."[8]

And countries such as Canada, the U.S., Russia, Venezuela, and Australia that rely heavily on the economic benefits of fossil fuel development—especially those that have failed to use the wealth from oil, coal, and gas to diversify the economy and put something away for hard times—have a difficult time shifting away from fossil fuels as quickly as necessary.

Countries that have started to shift show that it's entirely possible with political will. In 2008, Uruguay released a long-term energy policy to reduce reliance on imported oil and gas. It now gets 94.5 percent of its electricity from a mix of renewables—and 55 percent of its overall energy mix, including transport fuels—using mostly wind, solar, and biomass, with no nuclear or new hydro. According to the *Guardian* in 2015, Uruguay accomplished this by being a stable democracy that has never defaulted on its debts, with natural conditions that include lots of wind, sun, and agriculture for biomass, and strong public companies that work well with the private sector, along with a secure investment and regulatory environment.[9] The country has gone from being an energy importer to an exporter. Energy prices are lower relative to inflation than in the past, power supply is more reliable because of the diverse mix of energy sources, and the country of 3.4 million has cut its carbon footprint substantially.

Costa Rica managed to get all its electricity for ninety-four straight days in 2015 from a mix of hydropower (78 percent), geothermal (12 percent), and wind (10 percent), and plans to get all of its electricity from renewable sources by 2021.

The total share of renewable energy use in the European Union went from 8.7 percent in 2005 to 15.3 percent in 2014. The EU renewable energy directive's binding target calls for 20 percent final energy consumption and at least 10 percent of transport fuels from renewables by 2020, with individual national targets ranging from 10 percent in Malta to 49 percent in Sweden, according to the European Commission.[10] The target will rise to 27 percent by 2030. Sweden has already surpassed its 2020 goal, getting 52.1 percent of its energy from renewables in 2013. Bulgaria and Estonia have also surpassed their targets, Lithuania has met its target, and Romania and Italy are near their goals. Others, including the U.K., the Netherlands, France, and Ireland, are falling behind.[11]

Meanwhile, Denmark has set a goal of 70 percent renewables for energy by 2020 and 100 percent by 2050.[12] By compensating homeowners for losses in value from nearby wind farms, offering 20 percent of shares in a project to local communities, and burying cables underground, the government has increased public support. Denmark also plans to boost electric car use and connect batteries to the grid for storage, as well as using heat pumps for storage. Denmark's energy department says the economy grew by 78 percent since 1980, with energy consumption remaining about the same.

## Are Trade Deals Selling Out Our Future?

**WHEN INTERNATIONAL TRADE** agreements conflict with national and international environmental priorities, trade deals usually win. Many deals in effect or being negotiated allow corporations to sue at secret tribunals if national laws or policies, including environmental laws, are seen to negatively affect their profits and interests. This is because most international trade agreements include national treatment provisions that require goods to be treated equally whether produced locally or in another country.

And the massive increase of globalization and trade agreements has also caused a huge increase in greenhouse gas emissions. Transporting goods by ship, aircraft, rail, and road has increased massively over the past few decades of free trade agreements—and shows no signs of slowing. Because international shipping emissions aren't assigned to any one country, they aren't normally included in reductions targets. An International Maritime Organization report found that shipping alone accounted for 2.4 percent of global greenhouse gas emissions between 2007 and 2012,[13] and a European Parliament report found that could rise to 17 percent by 2050.[14]

Global trade has many benefits. For starters, it allows those of us who live through winter to eat fresh produce year round. And it provides economic benefits to farmers who grow that food. That could change as oil, the world's main transport fuel, becomes increasingly scarce, hard to obtain, and costly, but we'll be trading with other nations for the foreseeable future.

Because countries often have differing political and economic systems, agreements are needed to protect those invested in trade. Canada has signed numerous deals, from the North American Free Trade Agreement (NAFTA) to several Foreign Investment Promotion and Protection Agreements (FIPAS), and is subject to the rules of global trade bodies, such as the World Trade Organization (WTO).

Treaties, agreements, and organizations to help settle disputes may be necessary, but they often favour the interests of business over citizens. In 2014, Canada signed a thirty-one-year trade deal with China, a repressive and undemocratic country with state-owned corporations, which shows the need to be cautious.

Should we sign agreements if they subject our workers to unfair competition from lower-paid employees from investor nations, hinder our ability to protect the environment, or give foreign companies and governments excessive control over local policies and valuable resources? Under some agreements, basics like protecting the air, water, and land we all need for survival can become difficult and expensive.

One case could put Canada on the hook for $250 million.[15] Quebec has put a hold on hydraulic fracturing, or fracking, pending a study into the environmental impacts of blasting massive amounts of water, sand, and chemicals into the ground to fracture rock and release gas deposits. A U.S. resource company is suing Canada under 411 of NAFTA, claiming compensation for the moratorium's damage to its drilling interests.[16] Similar disputes have already cost Canada millions of dollars.[17]

Ontario also wants assurances that fracking is safe before it allows the practice.[18] That province is facing costs and hurdles because of another conflict between trade and the environment. Japan and the European Union filed a complaint with the WTO, claiming a requirement under the Ontario Green

Energy Act that wind and solar projects must use a set percentage of local materials is unfair.[19]

Many of the problems arise because of an investor-state arbitration mechanism, which is included in NAFTA, as well as the Canada-China FIPA, Canada–European Union Comprehensive Economic and Trade Agreement, and Trans-Pacific Partnership.[20] The mechanism allows foreign investors to bring claims before outside arbitrators if they believe their economic interests are being harmed by a nation's actions or policies. So economics trump national interests.

As Klein notes in *This Changes Everything*, "as of 2013, a full sixty out of 169 pending cases at the World Bank's dispute settlement tribunal had to do with the oil and gas or mining sectors, compared to a mere seven extraction cases throughout the entire 1980s and 1990s."[21] Many countries, including Australia, South Africa, India, and several in Latin America, have avoided signing deals that include the investor-state arbitration mechanism. This shows that governments don't have to stand for allowing trade agreements to get in the way of national priorities regarding the environment or other national interests. In Australia's case, the country recognized the pitfalls when tobacco companies, including Philip Morris, attempted to claim damages under a bilateral investment treaty after the federal government introduced a science-based law requiring cigarettes to be sold in plain, unappealing packages.

According to Australian National University law professor Thomas Faunce, Philip Morris then lobbied the U.S. government to include a similar mechanism in a new trade agreement it was negotiating with Australia. In an article for *Troy Media*, Faunce wrote that, with such a mechanism, the International Centre for the Settlement of Investment Disputes "would, in effect, become the final arbitrators on major Australian public policy questions concerning mineral royalties, fossil fuel and renewable energy, water, telecommunications, banking, agriculture and power."[22]

The thirty-one-year trade agreement between Canada and China is worrisome, with its fifteen-year opt-out clause (compared to just six months for NAFTA), but the inclusion of the mechanism in other agreements is also cause for concern. At the very least, Canada could be on the hook for millions or billions of dollars if its environmental, health, labour, or other policies were deemed to harm the interests of those investing in or trading with Canada. Similar trade deals between other countries carry the same risks.

## Infrastructure and the Fossil-Fuelled World

HAD THE WORLD started paying more attention to energy conservation and shifting from fossil fuels to cleaner energy sources when it was clear that burning coal, oil, and gas at such rapid rates was putting humanity at risk, we might have been able to engineer a more gradual, less disruptive shift. Now, we see the consequences of our bad habits hitting harder every day, yet we continue to build infrastructure, produce goods, and pursue activities that keep us locked into fossil fuels. From cars to pipelines to coal-fired power plants to fracking to offshore drilling and oil sands exploration, our practices don't reflect the kind of behaviour you'd expect from a species facing a crisis.

## WHERE'S CLIMATE IN THE DEBATE OVER TRANSPORTATION FUELS?

SOMETIMES, CLIMATE CHANGE isn't even raised as a factor in considering whether or not fossil fuel infrastructure projects should go ahead. Before Canada's government approved the Northern Gateway pipeline in 2014, the federal joint review panel examining the project refused to consider climate impacts or oil sands expansion in their decision, writing, "We did not consider that there was a sufficiently direct connection between the project and any particular existing or proposed oil sands development or other oil production activities to warrant consideration of the effects of these activities." As for climate change from burning the product, "These effects were outside our jurisdiction, and we did not consider them."[23] It's absurd. Those are the main reasons such a pipeline should not be constructed!

Fortunately, the project—which was to include a twinned pipeline to carry diluent from the coast to the tar sands to dilute bitumen that would then be carried back to the coast in another pipeline for export to world markets in supertankers—has not proceeded and is not likely to, but other pipeline projects are still in the works.

The arguments for building more pipelines so that more bitumen or fracked oil and gas can be shipped to world markets often ignore the very impacts that should be the priorities. Some focus solely on spill risks or economic factors. In its bid to triple shipping capacity from the Alberta tar sands to Burnaby on the southern British Columbia coast, in part by twinning its existing pipeline, energy giant Kinder Morgan attempted to downplay concerns about breaches and spills with economic arguments. "Pipeline spills can have both positive and negative effects on local and regional economies, both in the short- and long-term," a 2014 company submission to Canada's National Energy Board stated, adding, "spill response and cleanup creates business and employment opportunities for affected communities, regions, and cleanup service providers."[24]

In purely economic terms, the argument is true. The 2010 BP oil spill in the Gulf of Mexico is estimated to have added billions to the U.S. gross domestic product! But that's probably more of an argument for updating our economic systems, progress measurements, and ways of living to ones that don't depend on destroying everything the planet provides to keep us healthy and alive.

Yet another debate around transporting oil and gas is a red herring. Proponents say we need more pipelines because they offer a safer way to ship oil and gas than rail. It was something we heard a lot in 2013, after a train carrying fracked crude oil from North Dakota to a refinery in Saint John, New Brunswick, derailed in Lac-Mégantic, Quebec, caught fire, and caused explosions that destroyed much of the town and killed dozens of people, sending 5.6 million litres of oil into the ground, air, sewers, and the Chaudière River. Despite that and other rail disasters, others claim that leaks, high construction costs, opposition, and red tape surrounding pipelines are arguments in favour of using trains. But debating the best way to do something we shouldn't be doing in the first place is a sure way to end up in the wrong place.

To begin, both transportation methods come with significant risks. Shipping by rail leads to more accidents and spills, but pipeline leaks usually involve much larger volumes. One of the reasons we're seeing more train accidents involving fossil fuels is the incredible boom in moving these products by rail. According to the Association of American Railroads, train shipment of crude oil in the U.S. grew from 9,500 carloads in 2008 to 400,000 in 2013—a 4,111 percent increase in five years![25]

Ever-increasing rail accidents and pipeline leaks and spills aren't arguments for one or the other; they just show that rapidly increasing oil and gas development, and shipping ever greater amounts, by any method, will mean more accidents, spills, environmental damage, and often death. The answer is to step back from this reckless plunder and consider ways to reduce our fossil fuel use.

Slowing down and shifting away from continued fossil fuel exploitation, transportation, and use may result in losses of short-term jobs and economic opportunities the fossil fuel industry provides, but surely we can find better ways to keep people employed and the economy humming. Gambling, selling guns and drugs, and encouraging people to smoke all create jobs and economic benefits, too, but we rightly try to limit those activities when the harms outweigh the benefits. Pipelines and fossil fuel development are not great long-term job creators anyway, and pale in comparison to employment generated by the renewable energy sector. Although Canada is a major fossil fuel provider, a 2012 report by the Canadian Centre for Policy Alternatives showed that less than 1 percent of Canadian workers were employed in extraction and production of oil, coal, and natural gas.[26] The fact that the oil-producing province of Alberta was running consecutive budget deficits and experiencing increasing debt even before oil prices plummeted from more than $100 a barrel to well below $50 from summer 2014 to early the following year, and suffered even more as prices continued to drop, makes the economic arguments for increased exploitation and transportation suspect.

If we were to slow down development of the oil sands and other fossil fuel extraction projects, encourage conservation, and invest in clean energy technology, we could save money, ecosystems, and lives—and we'd still have valuable fossil fuel resources long into the future, perhaps until we've figured out ways to use them that aren't so wasteful. We wouldn't need to build more pipelines just to sell oil and gas as quickly as possible around the world. We wouldn't have to send so many unsafe rail tankers through wilderness areas and places people live. Of course, we must still improve rail safety and pipeline infrastructure for the oil and gas that we'll continue to ship for the foreseeable future, but we must also find ways to transport less.

Beyond dangers to the environment and human health, the worst risk from rapid expansion of oil sands, coal mines, and gas fields, and the infrastructure needed to transport the fuels, is the carbon emissions from burning their products—regardless of whether that happens in North America, China, or elsewhere. Many climate scientists and energy experts, including the International Energy Agency, agree that to have any chance of avoiding catastrophic climate change, we must leave up to 80 percent of our remaining fossil fuels in the ground.

The question isn't about whether to use rail or pipelines or trucks to transport fuels. It's about how to reduce our need for the fuels. Continuing to build fossil fuel infrastructure just locks us into continued fossil fuel use and the consequent pollution and climate change.

## PLANES, TRAINS, AND AUTOMOBILES

HENRY FORD'S INVENTION of the Model T car in the early twentieth century transformed America, and subsequently much of the rest of the world. As cars became cheaper to produce and buy, they were seen as a ticket to freedom and mobility. Infrastructure was built up around them, sparking economic growth, job creation, and international trade in products such as rubber and steel. Oil was plentiful, and fuel efficiency wasn't a concern because, before pollution and climate change impacts were known, gas sale profits were part of the economic engine driving prosperity.

Some even claim automobile and oil companies bought and dismantled streetcar and urban rail lines from the mid-1930s to the 1950s to sell more cars and oil. In the U.S. and Canada, and other parts of the world, cities were designed and built to accommodate cars rather than people, with urban sprawl, shopping malls, roads, freeways, bridges, and parking spaces everywhere. Many factors were involved in the development of car culture, but we now find ourselves in an era when much of our oil is burned to propel mostly single users in inefficient vehicles along infrastructure and through landscapes designed for drivers. That makes it much harder to shift away from private automobile use, especially when developing nations want the same benefits cars and trucks have provided in the industrialized world.

Cars and trucks are among the biggest contributors to the heat-trapping emissions that cause global warming. According to the Union of Concerned Scientists, "Collectively, cars and trucks account for nearly one-fifth of all U.S. emissions, emitting around 24 pounds [11 kilograms] of carbon dioxide and other global-warming gases for every gallon [3.8 litres] of gas. About 5 pounds [more than 2 kilograms] comes from the extraction, production, and delivery of the fuel, while the great bulk of heat-trapping emissions—more than 19 pounds per gallon [2 kilograms per litre]—comes right out of a car's tailpipe." When planes, trains, ships, and freight are included, transportation makes up about 30 percent of U.S. global warming emissions.[27]

The percentage of emissions from cars and trucks—and transportation in general—is higher in the U.S. than other countries, but private automobile use is increasing in some parts of the world, and continues to be a major factor in global warming, despite fuel-efficiency improvements and technological innovation in hybrid and electric vehicle production.

In China, which has embraced the private automobile, a massive traffic jam in 2010 stretched for one hundred kilometres and lasted almost two weeks!

Even with better fuel-efficiency standards, automobiles waste on average 85 percent of the energy from each tank of gas. And the useful energy goes to moving a vehicle that typically weighs ten to twenty times more than the passengers it carries. That translates to about 1 percent efficiency to move passengers. The average car in North America carries 1.5 people, which means that most cars on the road only have a driver in them. Is it really efficient to use more than 900 kilograms of metal to transport 90 kilograms of human?

Electric car technology is picking up, but it doesn't resolve all of the issues, especially as the electricity must come from somewhere, and in many places, that means coal-fired power plants. Car manufacturing is also energy intensive. And the infrastructure required to support the prevalence of individual-use cars—from roads to parking lots—can also contribute to climate change.

Climate change isn't the only reason to reduce reliance on cars, suvs, and trucks. Cars directly kill and hurt more people every year than most diseases, resulting in 1.5 million deaths and 78 million injuries needing medical care, according to the World Bank. Road injury is the eighth leading cause of death worldwide. Pollution from cars also causes acute and chronic health problems that often result in premature death—from heart disease and stroke to respiratory illness and lung cancer.

Environmental impacts of cars are also well known and wide ranging, including smog and oily runoff from roads that washes into rivers, lakes, and oceans, not to mention the green space sacrificed for infrastructure to sell, drive, fuel, and park them. In the U.S., there are eight parking spaces for every car. We also devote an incredible amount of real estate to our ever-expanding road systems, often at the expense of public spaces.

Despite fuel-efficiency improvements, emissions from vehicles have more than doubled since 1970, and will increase with rising car demand in countries such as China, India, and Brazil, according to the Intergovernmental Panel on Climate Change.

Although we can't stop using cars altogether, we can curtail their damage to people and the environment. We can reduce greenhouse gas emissions by cutting back on car use, choosing fuel-efficient vehicles, joining a car pool or sharing program, and reducing speed. At the policy level, we need increased investment in public transit and cycling and pedestrian infrastructure, stronger fuel-efficiency standards, reduced speed limits, higher gas taxes, and human-centric urban design.

Besides combatting pollution and climate change, reduced dependency on private automobiles will lead to healthier people, fewer deaths and injuries, and more livable cities with happier citizens.

## Doubt, Confusion, and Denial

ALTHOUGH THE PARIS Agreement showed that the world is ready to take global warming seriously, and that those who deny the evidence for human-caused climate change or the need to address it are in an ever-dwindling minority, denial, doubt, and confusion still play a major role in stalling much-needed action. A 2015 study published in *Global Environmental Change* found that as the evidence for anthropogenic climate change has grown, contrarian think tanks have stepped up efforts to deny and spread doubt and confusion about the science.[28] Authors of the study, "Text-Mining the Signals of Climate Change Doubt," looked at fifteen years of output from nineteen U.S. conservative think tanks, analyzing more than sixteen thousand documents up to 2013. Asked whether the denial may have abated since 2013, given that the two following years were the hottest on record and that almost all the world's nations agreed in late 2015 that the problem must be addressed, author Travis Coan of the University of Exeter told the *Guardian* in 2016 that after looking at the top fifty climate skeptic blogs though 2015, he and John Cook from the University of Queensland found that "the increase in science-related skepticism continues right through 2015," and that the end of science denial "might be wishful thinking."[29]

People associated with the fossil fuel industry still put a lot of money and effort into misleading campaigns to downplay the risks of their industry, and to convince people that shifting away from coal, oil, and gas would be too costly, difficult, or unnecessary. Often the money goes to organizations like those in the study mentioned above that publish reports, hold conferences, and spread information online and in publications through articles, interviews, letters, and comments. A 2015 study in the *Proceedings of the National Academy of Sciences of the USA* concluded that "organizations with corporate funding were more likely to have written and disseminated texts meant to polarize the climate change issue," and that "corporate funding influences the actual thematic content of these polarization efforts, and the discursive prevalence of that thematic content over time."[30]

The Heartland Institute in the U.S.; the Global Warming Policy Foundation in the U.K.; Ethical Oil and Friends of Science in Canada; and the International Climate Science Coalition, based in Canada but affiliated with similar organizations in Australia and New Zealand and with close ties to Heartland, are all part of the campaign to dismiss climate science. A number of industry-funded websites also promote fossil fuels at the expense of human life, including Climate Depot and Watts Up With That?

These secretive organizations rarely reveal funding sources; prey on the uninformed and ignorant; and blanket the media with opinion articles, letters to editors, and comments, often referring to misleading charts and graphs and bogus "studies" from organizations with names that imply they're scientific when they're anything but. They're assisted by a compliant news media and politicians who also receive fossil fuel industry funding. It's likely the people behind these organizations know they're lying but care more about making money and preserving the lopsided benefits of a polluting sunset industry than finding ways to contribute to human health, well-being, and survival.

Their misinformation is usually all over the map, changing regularly when one lie doesn't stick. The Heartland Institute's 2014 "International Climate Change Conference" in Las Vegas illustrated this desperate confusion. As Bloomberg noted, "Heartland's strategy seemed to be to throw many theories at the wall and see what stuck."[31] A who's who of fossil fuel industry supporters and antiscience shills variously argued that global warming is a myth; that it's happening but natural—a result of the sun or "Pacific Decadal Oscillation"; that it's happening, but we shouldn't worry about it; and that global cooling is the real problem. Misleading claims that $CO_2$ is a benign or even beneficial gas that stimulates plant growth and doesn't affect climate are also common.

The only common thread, Bloomberg reported, was the preponderance of attacks on and jokes about Al Gore: "It rarely took more than a minute or two before one punctuated the swirl of opaque and occasionally conflicting scientific theories."

Personal attacks and other logical fallacies are also common among deniers. Their lies are continually debunked, leaving them with no rational challenge to overwhelming scientific evidence that the world is warming and that humans are largely responsible. Comments under my columns about global warming include endless repetition of falsehoods such as "there's been no warming for 18 years" and "it's the sun," and references to "communist misanthropes," "libtard warmers," alarmists, and worse.

The few attempts at legitimate scientific study to dispute the prevailing theories of climate change have also often been contradictory, farfetched, and easily debunked. A 2015 study published in *Theoretical and Applied Climatology*, titled "Learning from Mistakes in Climate Research," examined some of the tiny percentage of scientific papers that reject anthropogenic climate change, attempting to replicate their results.[32]

In a *Guardian* article, coauthor Dana Nuccitelli said their study found "no cohesive, consistent alternative theory to human-caused global warming." Instead, "Some blame global warming on the sun, others on orbital cycles of other planets, others on ocean cycles, and so on."[33]

Nuccitelli and fellow researchers noted that about 97 percent of experts worldwide agree on a cohesive, science-based theory of global warming, but those who don't "are all over the map, even contradicting each other. The one thing they seem to have in common is methodological flaws like cherry picking, curve fitting, ignoring inconvenient data, and disregarding known physics."

It's astounding and tragic that, with all the evidence—from volumes of scientific research to the very real effects we are experiencing everywhere—some people stubbornly refuse to believe there's a problem worth addressing. Sadder still is that many of them are political leaders.

Those who argue that 7 billion people pumping massive amounts of greenhouse gases into the atmosphere aren't having a serious negative impact are out to lunch.

Fortunately, most thinking people don't buy the lies. People from all sectors and walks of life—religious, academic, business, political, activist, social justice, and citizenry—are calling for an urgent response to the greatest threat humanity faces. From Pope Francis and the Dalai Lama to Islamic scholars and Hindu, Sikh, and Jewish leaders; from Volvo, Ikea, and Apple to the International Monetary Fund, World Bank, and

World Health Organization; from every legitimate scientific academy and institution to enlightened political leaders—all have warned about the serious nature of global warming and the urgent need to do something about it. Polls and marches, demonstrations, and citizen initiatives show that people want action.

Why then, when we know that global warming is serious and that oil will run out, are we hell-bent on using it up as quickly as possible? Author and environmentalist Bill McKibben suggests a disturbing reason why people in the fossil fuel industry, the politicians they bankroll, and supportive deniers and organizations put profits ahead of the future of the planet and deny that climate change is a problem: the value of these immensely profitable industries "is largely based on fossil-fuel reserves that won't be burned if we ever take global warming seriously."

As McKibben notes, "ExxonMobil, year after year, pulls in more money than any company in history. Chevron's not far behind. Everyone in the business is swimming in money."[34] If they were to slow down production, or even admit that the future of humanity depends on leaving some of the resource in the ground, it would hurt their bottom lines.

Naomi Klein goes further in explaining why people continue to deny the existence of climate change, or at least the need to address it despite the overwhelming evidence that we can and must do something. Climate change threatens their very world view: "They have come to understand that as soon as they admit that climate change is real, they will lose the central ideological battle of our time—whether we need to plan and manage our societies to reflect our goals and values, or whether that task can be left to the magic of the market."[35]

And so we have politicians and industry shills using bogus talking points to discredit or silence those who are calling for sanity for the sake of our future. They falsely accuse us of wanting to shut down all industry and call us hypocrites because we are unable to completely disengage from the fossil fuel economy and infrastructure that humans have created.

## Real and Perceived Economic Barriers

**AS MICHAEL ZAMMIT** Cutajar, former executive secretary of the UN Framework Convention on Climate Change, told the *Guardian* in 2012: "Climate change is not just a distant threat but a present danger—its economic impact is already with us."[36] But we're to believe that corporate profits, ever-increasing growth, consumer culture, disposable products, and often-meaningless jobs to keep it all going are more important than the health and survival of humans and other species, and true long-term economic prosperity.

The failure of world leaders to act on the critical issue of global warming is often blamed on economic considerations. Over and over, we hear politicians say they can't spend our tax dollars on environmental protection when the economy is so fragile. Putting aside the absurdity of prioritizing and refusing to alter a human-created and adaptable tool like the economy over caring for everything that allows us to survive and be healthy, let's take a look at the economic reality.

A 2012 report, *Climate Vulnerability Monitor: A Guide to the Cold Calculus of a Hot Planet*, concluded that climate change was costing the world $1.2 trillion a year and eating up 1.6 percent of global gross domestic product (GDP), and rising.[37] It was also killing at least 400,000 people every year, mainly in developing countries. That's not counting the 4.5 million people a year who die from air pollution caused by burning fossil fuels.

The climate vulnerability report was compiled by fifty scientists, economists, and experts for the Europe-based non-governmental organization DARA and twenty countries that joined to form the Climate Vulnerable Forum.

As stated in the preface, it "challenges a conventional view: that global action on climate change is a cost to society. Instead, it enlightens our understanding of how tackling climate change through coordinated efforts between nations would actually produce much-needed benefits for all."

The report's authors also concluded that the challenges of global poverty and climate change "can be tackled simultaneously with the same policy framework that would shift our development path to a low-carbon footing," creating "jobs, investment opportunities, new possibilities for international cooperation and technological deployment to the benefit of all."

Although the researchers noted that adaptation must be part of any climate change strategy, they cautioned, "treating only the symptoms but not the cause of the climate crisis would result in spectacular economic losses for the world economy."

In failing to act on global warming, many leaders are putting jobs and economic prosperity at risk. It's suicidal, both economically and literally, to focus on the fossil fuel industry's limited short-term economic benefits at the expense of long-term prosperity, human health, and the natural systems, plants, and animals that make our well-being and survival possible. Those who refuse to take climate change seriously are subjecting us to enormous economic risks and forgoing the numerous benefits that solutions would bring.

The World Bank—hardly a radical organization—is behind one study that illustrates the folly of this way of thinking. Although it still views the problem and solutions through the lens of outmoded economic thinking, its report demolishes arguments made by those who argue it is not economically feasible to tackle climate change. "Climate change poses a severe risk to global economic stability," said World Bank Group president Jim Yong Kim in a news release, adding, "we believe it's possible to reduce emissions and deliver jobs and economic opportunity, while also cutting health care and energy costs."[38]

*Risky Business*, a report by prominent U.S. Republicans and Democrats, concludes, "The U.S. economy faces significant risks from unabated climate change," especially in coastal regions and agricultural areas.[39] We're making the same mistake with climate change we made leading to the economic meltdown of 2008, according to Henry Paulson, who served as treasury secretary under George W. Bush and sponsored the U.S. bipartisan report with former hedge fund executive Thomas Steyer and former New York mayor Michael Bloomberg. "But climate change is a more intractable problem," he argued in the *New York Times* in 2014. "That means the decisions we're making today—to continue along a path that's almost entirely carbon-dependent—are locking us in for long-term consequences that we will not be able to change but only adapt to, at enormous cost."[40]

One *Risky Business* author, former Clinton treasury secretary Robert Rubin, also warned about the economic risks of relying on "stranded assets"—resources that must stay in the ground if we are to avoid dangerous levels of climate change, including much of the bitumen in Canada's tar sands.[41]

Both studies recommend carbon pricing as one method to address the climate crisis, with the World Bank arguing for "regulations, taxes, and incentives to stimulate a shift to clean transportation, improved industrial energy efficiency, and more energy efficient buildings and appliances."[42]

In a commentary in Nature,[43] a multidisciplinary group of economists, scientists, and other experts called for a moratorium on all oil sands expansion and transportation projects such as pipelines because of what they described in a news release as the "failure to adequately address carbon emissions or the cumulative effect of multiple projects."[44] They want "Canada and the United States to develop a joint North American road map for energy development that recognizes the true social and environmental costs of infrastructure projects as well as account for national and international commitments to reduce carbon emissions."

Those who fear or reject change are running out of excuses as humanity runs out of time. Pitting the natural environment against the human-invented economy and placing higher value on the latter is foolish. These reports show it's time to consign that false dichotomy to the same dustbin as other debunked and discredited rubbish spread by those who profit from sowing doubt and confusion about global warming.

"Climate inaction inflicts costs that escalate every day," World Bank Group vice-president Rachel Kyte said, adding its study "makes the case for actions that save lives, create jobs, grow economies and, at the same time, slow the rate of climate change. We place ourselves and our children at peril if we ignore these opportunities."

It shouldn't be surprising to realize that using finite resources in a wasteful manner and at a pace much faster than the earth's ability to replenish them is economic folly. Nor should we be surprised to learn that polluting, damaging, and destroying the natural systems that keep us alive and healthy will not be good for our long-term prosperity, economic or otherwise.

Rapid population growth and technological innovation, combined with our lack of understanding about how the natural systems of which we are a part work and interconnect, have created a mess. We have altered the physical, chemical, and biological properties of the planet on a geological scale.

We must learn to shift our ways of thinking. We have to stop using so many disposable plastic items and fossil fuels. We need to conserve resources and energy and stop being so destructive. The DARA report offers many recommendations for governments, policy-makers, civil society, the private sector, and the international development and humanitarian communities.

Its advice for communicators applies to all of us: question received wisdom, recognize awareness of risks as an opportunity, and take a stand. Economies must function to serve people, not just shortsighted and often-destructive corporate interests. That means rethinking a system that relies on constant growth. After all, the planet and its resources are finite.

American journalist Charles Bowden once wrote, "Imagine the problem is that we cannot imagine a future where we possess less but are more."[45] Not being able to even imagine an economy without continual growth is a profound failure.

But most current economic systems favour short-term profits at the expense of long-term health and survival. Part of the problem is that our current economic systems are relatively new. In the 1930s and '40s, world leaders had to address unemployment and underproduction. Many of our economic measures were developed when natural capital (the benefits that nature provides) was plentiful but built capital (buildings, machinery, infrastructure) was not. In providing more manufactured goods and services, we developed a blind spot to the economic importance of natural systems. Labour, built, and financial capital are typically considered as the primary factors of production for economic development. Land and natural systems have seldom been included.

With growing human populations and profit-driven, consumer-based economics, more land is being eaten up by development, habitat is being destroyed and degraded, and resources are being exploited at unsustainable levels. Natural capital is disappearing.

We're also left with rapidly growing income inequality. In 2010, the world's richest 388 people had more wealth than the poorest 50 percent. In 2015, the gap grew to the point that sixty-two billionaires owned as much wealth as the poorest half. The late U.S. Supreme Court justice Louis Brandeis once noted, "We can have democracy in this country or we can have great wealth concentrated in the hands of a few, but we cannot have both."[46]

Growing inequality leads to increased human misery, and the idea that economic growth offers greater happiness to more people is a myth. A cornerstone of our current economy, consuming goods, may give us fleeting pleasure, but it isn't making us happier. Studies show the pleasure derived from food, sex, exercise, and time with loved ones or doing meaningful work takes much longer to fade. Worse, consuming stuff is not only addictive, it also feeds rivalry and societal over-consumption.

The Canadian Index of Wellbeing uncovered some troubling truths about the connection between the economy and well-being.[47] When Canada's economy was thriving, Canadians saw only modest improvements in their overall quality of life, but when the economy faltered, our well-being took a disproportionate step backward. Why are we so reluctant to talk about how we can get out of this cycle of endless buying and unsatisfying consumption by considering steady-state economies (which are stable or only fluctuate mildly and do not exceed ecological limits) or even degrowth alternatives?

The economy is a human construct, not a force of nature like entropy, gravity, or the speed of light, or our biological makeup. It makes no sense to elevate the economy above the things that keep us alive. This economic system is built on exploiting raw materials from the biosphere and dumping the waste back into the biosphere. And conventional economics dismisses all the "services" that nature performs to keep the planet habitable for animals like us as "externalities." As long as economic considerations trump all other factors in our decisions, we will never work our way out of the problems we've created.

We often describe the triple bottom line—society, economy, and environment—as three intersecting circles of equal size. This is nonsense. The reality is that the largest circle should represent the biosphere. Within that, we have 30 million species, including ours, that depend on it. Within the biosphere circle should be a much smaller circle, which is human society, and within that should be an even smaller circle, the economy. Neither of the inner circles should grow large enough to intersect with the bigger ones, but that's what's happening now as human societies and the economy hit their limits.

We also draw lines around property, cities, provinces, and countries. We take these so seriously that we are willing to fight and die to protect those borders. But nature pays no attention to human boundaries. Air, water, soil that blows across continents and oceans, migrating fish, birds and mammals, and windblown seeds cannot be managed within human strictures, yet international climate discussions centre on countries that are divided into rich and poor. In science-fiction movies where an alien from outer space attacks and kills humans, national differences disappear as we join forces to fight a common enemy. That is what we have to tap into to meet the climate crisis.

Nature is our home. Nature provides our most fundamental needs. Nature dictates limits. If we are striving for a truly sustainable future, we have to subordinate our activities to the limits that come from nature. We know how much carbon dioxide all the green things in the oceans and on land can reabsorb, and we know we are exceeding those limits. That's why carbon is building up in the atmosphere. So our goal is clear. All of humanity must find a way to keep emissions below the limits imposed by the biosphere. The only equitable course is to determine the acceptable level of emissions on a global per capita basis. Those who fall below the line should be compensated for their small carbon footprint, whereas those who are far above should be assessed accordingly. But the economy must be aligned with the limits imposed by the biosphere, not above them.

Despite the failure to imagine a better way, we may finally be seeing a change in course. The climate crisis is creating a global consciousness shift, with hundreds of thousands marching to demand change, and people from business, government, religion, non-governmental organizations, and more warning that economic growth and technology can't continue to trump all other concerns. Throughout Europe, North America, and beyond, support is growing for confronting income and wealth inequality and, at the same time, the growing climate crisis.

## Does Population Growth Make Global Warming Inevitable?

THE WORLD'S HUMAN population has grown exponentially, from 1 billion in 1800 to about 2 billion in 1927 and more than 7 billion today. Although growth is slowing somewhat, it's expected to hit 9 billion by 2040. In my lifetime alone, the human population has more than tripled. (I know I've contributed to the boom.) That puts enormous pressure on the environment and climate. More people means more resources used, more fossil fuels burned, and more carbon-intensive agriculture practised.

Supporting more people on a finite planet with finite resources is a serious challenge. But in a world where hunger and obesity are both epidemic, reproduction rates can't be the only problem. When we look at issues that are often blamed on overpopulation, we see that over-consumption by the most privileged is a greater factor in rampant environmental destruction and resource depletion.

I once asked the great ecologist E.O. Wilson how many people the planet could sustain indefinitely. He responded, "If you want to live like North Americans, 200 million." North Americans, Europeans, Japanese, and Australians, who make up 20 percent of the world's population, are consuming more than 80 percent of the world's resources. We are the major predators and despoilers of the planet, so we blame the problem on overpopulation. Most environmental devastation is not directly caused by individuals or households but by corporations driven more by profits than human needs, but it still takes people to keep the system going.

The nonprofit organization Global Footprint Network calculated the area of land and water the world's human population needs to produce the resources it consumes and to absorb carbon dioxide emissions.[48] If it takes a year or less for nature to regenerate the amount we use in a year, that's sustainable. But they found it takes 1.5 years to replace what we take in a year. That means we are using up our basic biological capital rather than living on the interest, and this has been going on since the 1980s.

As people in developing countries demand more of the bounty and products we take for granted, environmental impacts are bound to increase, including climate change. The best way to confront these problems is to reduce waste and consumption, find cleaner energy sources, and support other countries in finding ways to develop that are more sustainable than the ways we've employed—to learn from our mistakes. Stabilizing or decreasing population growth will help, but research shows it's not the biggest factor. A United Nations report, *State of World Population 2011*, concludes that even zero population growth won't have a huge impact on global warming.[49]

But, just as it's absurd to rely on economies based on constant growth on a finite planet, it can't be sustainable to have a human population that continues to increase exponentially. The good news is that population growth is coming down. According to the UN report, the average number of children per woman went from 6 to 2.5 from about 1950 to 2011. Research shows the best way to stabilize and reduce population growth is through greater protection and respect for women's rights, better access to birth control, widespread education about sex and reproduction, and redistribution of wealth.

But wealthy conservatives who overwhelmingly identify population growth as the biggest problem are often the same people who oppose measures that may slow the rate of growth. This has been especially true in the U.S., where many corpo rate leaders and the politicians who support them fight against environmental protection and against sex education and better access to birth control, not to mention wealth redistribution.

Population, environmental, and social justice issues are inextricably linked. Giving women more rights over their own bodies, providing equal opportunity for them to participate in society, and making education and contraception widely available will help stabilize population growth and create numerous other benefits. Reducing economic disparity—between rich and poor individuals and nations—will lead to better allocation of resources. Stabilizing population growth is clearly an important factor in reducing the human impact on Earth's natural systems, including climate, but it will take more than that to confront serious environmental problems like global warming.

# PART 2

## The Solutions

IN A POWERFUL example of hope for humanity, the 2015 Paris Agreement was an acknowledgement by world leaders and experts that global warming is a global crisis—one that goes beyond national borders and priorities—and that it will therefore require a massive worldwide effort to resolve. But the fact that it took until 2015 for the world's nations to come up with what is essentially an inadequate agreement to confront this global emergency shows just how difficult it is to steer humanity onto a sustainable course. That some people view this kind of international agreement as some kind of nefarious plot to implement a world government and transfer wealth rather than acceptance of our shared humanity and need for a healthy planet indicates how important it is for the rest of us to do our part.

The climate crisis won't be resolved by a nonbinding agreement alone that only gets the world halfway to the emissions reductions scientists agree are necessary to avert catastrophic warming. Nor will trading in your car for a bicycle or bus pass resolve it. Carbon taxes or cap and trade alone won't stop people and corporations from polluting the air and emitting $CO_2$.

Reducing your meat consumption or even becoming a vegetarian or vegan won't stop the world from warming. But all of
these will help, and all will contribute to the necessary resolu-
tion of this urgent and worsening threat to humanity.

Although we've known for decades, if not more, that our
wasteful ways spell disaster for the planet's ability to support
human and other life, and that solutions exist, many factors
have led us to continue with business as usual. Fossil fuels are
cheap and plentiful, found in one form or another all over the
world, and the technologies that allow us to use them for our
benefit are pretty basic: we just have to burn them and they
release huge amounts of energy captured from the sun eons
ago through photosynthesis and concentrated through geological forces over millennia! They're stable and relatively easy
to transport. They've made life a whole lot easier for rapidly
growing human populations—increasing our mobility, lighting
our homes and cities, facilitating rapid technological advances,
rationalizing food systems, reducing toil, and so much more.

Shifting to cleaner energy sources would have been easier
if we had started earlier, when we started realizing that our
path wasn't sustainable. Had the world heeded scientists like
Svante Arrhenius in the late nineteenth century or Mikhail
Budyko more recently—when they warned that excessive
burning of fossil fuels could upset climate systems and trigger
feedback cycles—and taken a precautionary approach before
our fossil fuel use got out of hand, we'd probably be on a more
prosperous and healthy road by now, and at least on the way to
resolving this crisis if not outright averting it. If we'd at least
listened to climate scientists like James Hansen, who testified
in Washington in 1988 that human-caused global warming
had kicked in and that it spelled real trouble for humanity,

we'd be better off than we are. But when people have it easy, as do many in the industrialized world, they're reluctant to change course. When such massive profits are being made from exploiting and using fossil fuels as quickly and often wastefully as possible, the challenge becomes even more formidable. Along with the benefits that fossil fuels have given us, they have also helped drive an economic system that encourages waste and consumption for the sake of increasing profits and creating more work.

There has been progress over the years, at national and subnational levels, and among forward-thinking corporations and organizations. Some, like the commitment by countries including Denmark and Germany to reduce dependence on fossil fuels after the 1973 Arab oil embargo, were in response to markets rather than the climate crisis, but their efforts positioned them well as evidence for climate change mounted. More recently, people on the frontlines of climate change, such as Pacific Islanders and Inuit, have warned of the changes they're experiencing. The insurance industry and a number of corporations have called for action, with some, like Tesla, designing solutions. But many in the media and government continue to downplay the problem.

Although people have been employing solutions—developing ways to conserve energy, creating and employing clean energy technologies, reducing waste, and taking steps to reduce their personal footprint—it hasn't been enough to avert the crisis we now face. Even if we somehow found ways to immediately stop burning fossil fuels and emitting greenhouse gases, we wouldn't stop climate change. The emissions we've already spewed into the atmosphere aren't going to be absorbed immediately, and feedback cycles have already been triggered. We also have to address the climate impacts of our food and agricultural systems. This means we have to consider and employ every solution we have available, and we have to continually find new solutions, from small changes in our immediate lives to massive global shifts in technology, practices, and ways of thinking and acting. We also need to stabilize population growth, in large part by promoting women's rights. And we have to find ways to adapt to the changes we can no longer prevent.

I've been astounded by the lack of response over the years, but I'll go out on a limb and suggest a shift is taking place. Although we may not recognize its significance without the benefit of hindsight, we appear to be in the early stages of something huge. Even some news outlets are shifting. The *Guardian* decided in early 2015 to increase its coverage of climate change, going so far as to encourage divestment from the fossil fuel industry. The *New York Times* decided to use the more accurate term "denier" rather than "skeptic" to refer to those who reject the overwhelming evidence for human-caused climate change.

People power is another sign of the growing shift: 400,000 at the largest climate march in history in New York in September 2015, with 2,646 simultaneous marches in 162 countries; an unprecedented gathering of 25,000 in Quebec City in advance of a Canadian premiers' climate change summit in April 2015; and more than 10,000 in Toronto (including me) on July 5 the same year for the March for Jobs, Justice and the Climate in advance of the Climate Summit of the Americas. When Pope Francis reached beyond the world's 1.2 billion Catholics to call for action on climate change in his 2015 encyclical, other religious leaders and organizations endorsed his message, including the Dalai Lama, the Islamic Society of North America, an influential group of Jewish rabbis, and the Church of England.

Beyond visible evidence of the increasing willingness to meet the challenge of global warming, one of the biggest signs of a shift has been the almost unnoticed but spectacular increase in renewable energy investment in countries such as the U.S., Brazil, and China.

Even well-known deniers, including U.S. oil billionaire Charles Koch, now admit climate change is real and caused in part by $CO_2$ emissions. But they argue it isn't and won't be dangerous, so we shouldn't worry. Most people are smart enough to see through their constantly changing, antiscience, pro-fossil fuel propaganda, though, and are demanding government and industry action. We're also seeing significant changes in the corporate sector. The movement to divest from fossil fuels is growing quickly, and businesses are increasingly integrating positive environmental performance into their operations. Funds that have divested from fossil fuels have outperformed those that haven't, a trend expected to continue.

It's easy for governments and industry to prioritize corporate profits and short-term gain over the best interests of complacent citizens. But when enough people demand action, take to the streets, write to business, political, and religious leaders, and talk to friends and family, change starts happening. We never know how big it will be until it's occurred—but this time, it looks like it could be monumental! Let's hope so. We need a massive social movement to confront this enormous challenge.

In considering the range of possible solutions, we have to evaluate which are likely to work best and which may be counterproductive. That's not an easy task. In an article for the *Huffington Post*, sustainability writer and Liology Institute founder Jeremy Lent offers three ways to evaluate proposed large climate solutions:[1]

1.  Does it push political power up or down the pyramid? Lent notes that some technologies concentrate power with governments and the very wealthy, such as nuclear power, which requires "massive centralized investment along with extreme security," and geoengineering, which is often proposed or developed "by small, elite groups of technical experts, usually funded by corporate interests," who "frequently envisage forcing a global experiment on the entire world, regardless of whether there is a consensus supporting such an approach." Solar power, however, can be affordable even for people with limited resources.

2.  How does it treat the earth? Lent writes, "If a proposed climate solution treats nature in the same way that has brought our civilization to the precipice we're now facing, there's a good chance it will lead to further environmental devastation. In contrast, proposals that encourage a participatory, engaged relationship between humans and the natural world promise a more sustainable future for all."

3.  What are its cascading effects? Solutions that work with nature tend to be better than those that seek to alter conditions on a large scale, which often come with unintended consequences.

In the following four chapters, we'll look at some of the existing and proposed solutions and examine which are most likely to get us where we need to go, which may help us live with the changes we've already triggered, and which may hinder us from resolving the crisis. We'll look at what you can do as an individual, what communities can do, and what industries and government must do. We're going to see the effects of climate change continue, with consequences getting worse, for some time, but if we all do our part, we can hope to live with the inevitable changes and to avert complete catastrophe.

Chapter 4

# PERSONAL SOLUTIONS

I**T'S DIFFICULT NOT** to get discouraged by the endless
stream of horrific world events—the Syrian war
and refugee crisis; violent outbreaks in Beirut, Paris,
Burundi, the U.S., and so many other places; and the ongoing climate catastrophe. But responses to these tragedies and
disasters offer hope. It's becoming increasingly clear that when
those who believe in protecting people and the planet; treating each other with fairness, respect, and kindness; and seeking
solutions stand up, speak out, and act for what is right and just,
we will be heard.

As Syria descended deeper into chaos during 2015, people in many wealthy nations called for blocking refugees.
But many more opened their hearts, homes, and wallets and
showed compassion. Governments responded by opening
doors to people who had lost everything, including family and
friends, to flee death and destruction. Shootings and the inevitable absurd arguments against gun control continued south
of the border, but many people, including President Barack
Obama, rallied for an end to the insanity. And although the

2016 U.S. presidential race was mired in bigotry, ignorance, and a dumbfounding rejection of climate science, many U.S. citizens, including political candidates, spoke out for a positive approach more aligned with America's professed values. And in 2015, voters in Canada and elsewhere rejected fear-based election campaigns that promoted continued reliance on climate-altering coal, oil, and gas.

The fossil fuel industry and its supporters continued to sow doubt and confusion about the overwhelming evidence for human-caused climate change and to rail against solutions, but many more people marched; signed petitions; sent letters; talked to friends and family; demanded action from political, religious, and business leaders; and started innovating and implementing solutions. The public appetite for a constructive approach to global warming led Canada to shift course in 2015, taking global warming seriously enough to make positive contributions at the Paris climate conference in December.

If we want to heal this world we have so badly damaged, we must do all we can. Although many necessary and profound changes must come from governments, industry, and other institutions, we can all do our part. For the climate, we can conserve energy, eat less meat, drive less, improve energy efficiency in our homes and businesses, and continue to stand up and speak out.

Those who fear and reject change have always been and always will be with us. They've argued ending slavery would destroy the economy; they've claimed putting people on the moon would be impossible; they've rejected ending South Africa's apartheid system; they've said the Berlin Wall wouldn't come down. With today's technological and communications advances, everyone with access to the Internet can be heard.

That's good, but people who fear they have something to lose often speak loudest, and in the greatest numbers. I tell people at the David Suzuki Foundation, "Don't read the comments!" It's often disheartening to see online discourse sink to such irrational and often idiotic depths.

But many comments and efforts to stall or block necessary progress arise from fear. People who are afraid that change might remove or diminish their privilege—real or imagined—often do or say anything to block it. Unfortunately, those who benefit most from privilege or the status quo, even if only in the short term, often stoke those fears and uncertainties, taking advantage of and manipulating the frightened and ignorant for political or economic gain.

That's not to say people must always agree. But racism, sexism, homophobia, religious prejudice, the denial of climate science and solutions, and blindness to the need for gun control are irrational. We can and must speak louder than those who would keep us on a destructive path despite the overwhelming evidence that it's past time to shift course. Events in 2015 taught us that when those of us who care about humanity and the planet's future stand up and speak out, we can make this small blue world and its miraculous life and natural systems a better place for all.

Climate change has reached a point where resolving it will require serious commitments and action from the top—from governments and industry. The 2015 Paris Agreement at least showed they are getting serious about commitments. But individual actions are also crucial. Sometimes what we do in our lives seems inconsequential when it comes to overwhelming issues like global warming, but all actions, no matter how small, add up to a larger whole. And as more people join the growing global movement to demand change, those who are supposed to represent our interests must listen.

Every action we take has ripple effects, raising awareness and encouraging others to join us. Although powerful bodies such as governments, industry, and international organizations bear the burden of responsibility, we can't leave it all up to them. The David Suzuki Foundation lists the top ten things individuals can do to help slow or stop climate change: green your commute, be energy efficient, choose renewable power, eat wisely, trim your waste, let polluters pay, fly less, get informed, get involved, and support and donate to organizations working to stop climate change. Most importantly, we need to get informed, get active, and join together to demand change. In this chapter, we'll look at some personal solutions in more detail.

## Green Your Commute

TRANSPORTATION (INCLUDING FREIGHT transport) is a major contributor to greenhouse gas emissions, accounting for a quarter or more of emissions in most industrialized countries, with about half of that from personal automobiles. If you live in a city, it doesn't take much to see the enormous impact of cars, SUVs, and trucks. Anything you can do to reduce the amount you drive, or to reduce the amount of fuel you use when you must drive, helps cut emissions. The advantages of getting out of your car go beyond reducing pollution and greenhouse gas emissions. Cycling and walking are good for your health, even if it's just walking to and from the bus stop or metro station. When we're not isolated in our cars, we connect more with others and the world around us. Driving, especially during rush hour, can be stressful. Walking, cycling, or reading or listening to music while riding public transit can reduce stress and make people happier and more productive at work and in their personal lives. And driving less saves money that you would have to spend on fuel, maintenance, and parking. Getting rid of your car altogether can save even more, considering the costs of a vehicle and insurance.

## RIDE A BIKE

ONE TRANSPORTATION METHOD is so efficient, beneficial, and simple that it may be the best thing we've ever invented. The "modern" version of the bicycle with pedals and cranks was invented by French carriage-maker Ernest Michaux in 1861. It's come a long way since then, but whether it's a high-tech racing bike or a one-gear street cruiser, the bike is still a marvel of ingenuity. In fact, it may well be the most efficient form of transportation yet invented.

The best part of the bike is that the rider is the engine. The fuel is what you eat and drink. Putting the human engine together with the gears, wheels, and frame of a bike gives you a mode of transportation that uses less energy even than walking. As for our most popular method of getting around, the automobile, there's no comparison. According to the Worldwatch Institute, a bicycle needs 35 calories per passenger mile, whereas a car uses 1,860. Buses and trains are somewhere in between.[1]

Bicycling is a fast, efficient, cost effective, and healthy way to get around. Of course it's not always possible to ride, and it's not for everyone. Depending on where you live, weather and lack of cycling infrastructure can be barriers to cyclists. And some people with physical mobility issues aren't able to take up cycling. But with proper clothing and gear, many people can ride for most of the year in urban centres. And the money saved from not driving is often enough to pay for public transit or taxis on days when cycling isn't possible. We could be doing much more in most cities than we are. In Canada, only about 1 percent of trips are made by bike (though Vancouver is higher, at about 4 percent), whereas in many parts of Europe, the number is more than 30 percent. In Amsterdam, 38 percent of trips are made by bike, thanks to pro-cycling policies adopted since the 1970s.

The number of trips made on bicycles has been steadily increasing in the U.S. as well. According to the League of American Bicyclists, bicycle commuting increased by 62 percent from 2000 to 2013 in the U.S. In bike-friendly communities, where infrastructure has been put in place to encourage cycling, the rate was 105 percent! Some cities, such as Washington, D.C., and Portland, Oregon, have seen huge increases, at 498 and 408 percent respectively.[2] Still, according to the U.S. Census Bureau, the percentage of people who commuted to work by bicycle in the country's fifty largest cities only rose from 0.6 percent to 1 percent from 2000 to 2008–12. In Portland, the rate for the 2008–12 census was 6.1 percent, proving that good infrastructure can make a difference.[3]

As more people take up cycling, it also becomes safer. Although, those who worry about the safety of cycling might be interested in a British Medical Association study that found the health risks of inactivity are twenty times greater than the risks from cycling.

For employers, the benefits of encouraging cycling are numerous. A Dutch study found that people who cycle to work take fewer sick days, and research has shown they are generally happier and less stressed. Cyclists can also avoid traffic jams and are not as likely to be late for work. And bike lockups cost far less than parking facilities.

Cleaner air, reduced congestion, safer streets, and lower noise levels are just a few of the benefits of getting more people out of their cars and onto bikes. People who cycle also become fitter, leading to lower health care spending.

Getting out of your car and onto a bike whenever possible makes sense. Because good infrastructure makes cycling safer, easier, and more efficient, advocating for bike lanes and other incentives for cycling is also important, especially given the resistance that drivers and others often put up. Most of the arguments against bike lanes and other infrastructure are absurd. Consider this: We have wide roads everywhere to accommodate cars, most of which carry only one person. On either side of many of those roads, we have pedestrian sidewalks. In most large urban areas, we also have bus lanes and transit systems such as subways and rapid transit. When cyclists ride on roads, drivers often get annoyed. If they ride on sidewalks, pedestrians rightly get angry.

In many North American cities, including Vancouver, where I live, commuters scream bloody murder if it takes them an extra two minutes to get to their destination by car. The reality is that drivers are slowed more by increases in car traffic than by bike lanes. A study by Stantec Consulting Ltd. found that traffic delays because of bike lanes in Vancouver were mostly imagined.[4] Drivers who were surveyed thought it took them five minutes longer to travel along a street with a new bike lane. But the study showed that it actually took from five seconds less to just a minute and thirty-seven seconds more.

Even if bike lanes did slow traffic, that might not be such a bad thing. In some European cities, planners are finding that making life more difficult for drivers while providing incentives for people to take transit, walk, or cycle creates numerous benefits, from reducing pollution and smog-related health problems to cutting greenhouse gas emissions and making cities safer and friendlier.

And it's good for business. Where streets were closed to cars in Zurich, Switzerland, to make way for cycling and pedestrian infrastructure, store owners worried about losing business, but the opposite happened—pedestrian traffic increased 30 to 40 percent, bringing more people into stores and businesses. In the long run, most cities that have improved cycling and pedestrian infrastructure have seen benefits for area businesses.

### WALK FOR YOUR HEALTH AND THE PLANET

WALKING MAY NOT be as quick as cycling, but if you live near enough to your workplace or other places you have to go, it's an enjoyable and healthy way to get around. As the American comedian Steven Wright says, "Everywhere is within walking distance if you have the time." Walking helps you maintain a healthy weight, strengthens bones and muscles, improves balance and coordination, helps you sleep better, lifts your mood, and helps prevent or manage conditions including heart disease, high blood pressure, and type 2 diabetes.

According to the Arthritis Foundation, studies of postmenopausal women have found that just thirty minutes of walking a day reduced the risk of hip fractures by 40 percent and reduced the risk of stroke by 20 to 40 percent, depending on their pace.[5] Research also shows it can reduce mental decline and Alzheimer's disease risk, as well as incidence of disability in older people. Fortunately, most cities offer sidewalks and other infrastructure for pedestrians, but urban design can also go a long way to encouraging people to walk. Safer streets, better lighting, reduced urban sprawl, park and green spaces, compact shopping and business areas, public art, and pedestrian crossing lights all encourage people to walk rather than drive.

In Zurich, planners have added more traffic lights, including some that transit operators can change in their favour, increased the time of red lights and decreased the greens, removed pedestrian underpasses, slowed speed limits, reduced parking, and banned cars from many streets. "Our goal is to reconquer public space for pedestrians, not to make it easy for drivers," chief traffic planner Andy Fellmann told the *New York Times*.[6] He also noted that a person in a car takes up 150 cubic metres of urban space in Zurich, whereas a pedestrian takes up 3.

Walking is the perfect example of the idea that what we do for ourselves can also benefit the planet. Even taking public transit means people walk more, as they have to get to and from transit stops—which contributes to better physical and mental health.

## TAKE PUBLIC TRANSIT

IF YOU WANT to ditch the car and walking or cycling aren't viable options, public transit is often a good way to get around. Buses, trains, and rapid transit lines take up less space overall than cars, reduce congestion and pollution, emit fewer greenhouse gas emissions, and are often faster than cars. And you don't have to worry about driving, so you're left with time for reading, listening to music, relaxing, or even checking your email or Facebook! But not every city offers reliable public transit.

Great cities are built for people, not cars. New York City is world class not just because it's a driver of global finance and a hotbed of cultural innovation. It's also known for its green spaces, such as Central Park and the award-winning High Line. San Francisco is celebrated for its narrow streets, compact lots, and historic buildings. These contribute to the city's old-world charm, but they're also the building blocks of a more sustainable urban form. They facilitate densification and decrease the cost of energy and transportation for businesses while improving walkability. Cities such as Vancouver, San Francisco, Portland, Seattle, and New York City, which have consistently ranked among the most livable cities on the continent, also take the environment into account for planning decisions. They all have world-class public transit systems that move residents in a safe, affordable, and sustainable way.

A number of European cities have matched disincentives to drive with improved public transit, which even benefits vehicle drivers. With fewer cars and reduced gridlock, those who must use automobiles—including service and emergency-response vehicles and taxis—have an easier time getting around.

Many cities haven't invested in public transportation, to their detriment. A lack of public transit options means more people drive, which means more pollution and more greenhouse gas emissions. With more cars on the road, you also get more traffic congestion, which causes even more pollution and greenhouse gas emissions, not to mention increased frustration as people sit and fume in traffic.

People who live in cities with good public transit are fortunate, because not only is riding buses, trains, and the subway much better for the climate than driving, it's also better for your health. Research on the relationship between health and transit use in Metro Vancouver by University of British Columbia urban planning and public health professor

Lawrence Frank and two health authorities revealed that residents of areas with above average public transportation use are 26 percent less likely to be obese and 49 percent more likely to walk for at least thirty minutes a day than people living in low transit use areas.[7]

There's growing recognition that prioritizing transit is crucial to moving a region forward. Since the 1970s, Curitiba, Brazil, a city of 1.9 million, has invested billions in its bus rapid transit (BRT) network. There, public transportation is fully integrated into planning decisions. High-density hubs with shopping centres and office buildings are located within walking distance of transit stations, and commuters have access to a fleet of more than two thousand modern, low-emission buses, servicing 390 routes that crisscross the city and connect it to surrounding communities. Eighty-five percent of Curitiba's residents use the BRT system, which has reduced car trips by a whopping 27 million a year.

But, as with bike lanes, and anything else that might cut into profits for the car and oil industry, transit improvements are often met with opposition. It's been going on for a long time.

In the U.S., researchers found that brothers Charles and David Koch, who run Koch Industries, the second-largest privately owned company in the U.S., have backed many anti-transit initiatives in the U.S., in cities and states including Nashville, Indianapolis, Boston, Virginia, Florida, and Los Angeles.[8] They've also given close to US$70 million to climate change denial front groups, some of which they helped start, including Americans for Prosperity, founded by David Koch and a major force behind the Tea Party movement.[9]

Through their companies, the Kochs are the largest U.S. leaseholder in the Alberta oil sands.[10] They've provided funding to Canada's pro-oil Fraser Institute and are known to fuel the Agenda 21 conspiracy theory, which claims a 1992 UN nonbinding sustainable development proposal is a plot to remove property rights and other freedoms.[11] They spend large amounts of money on campaigns to discredit climate science and the need to reduce greenhouse gases, and they fund sympathetic politicians.

In early 2015, fifty U.S. anti-government and pro-oil groups—including some tied to the Kochs and the pro-oil, pro-tobacco Heartland Institute—sent Congress a letter opposing a gas tax increase that would help fund public transit, in part because "Washington continues to spend federal dollars on projects that have nothing to do with roads like bike paths and transit."[12] The letter says, "Transportation infrastructure has a spending problem, not a revenue problem," an argument similar to one used by opponents to a 2015 transportation plan in Metro Vancouver. Vancouver's anti-transit campaign was led by the Canadian Taxpayers Federation—a group that doesn't reveal its funding sources and is on record as denying the existence of human-caused climate change[13]—along with Hamish Marshall, a conservative strategist with ties to Ethical Oil, an organization that promotes the oil sands.

American and Canadian transit opponents paint themselves as populist supporters of the common people, a tactic also used against carbon pricing. Marshall told *Business in Vancouver* in 2015, "I love the idea of working on a campaign where we can stand up for the little guy."[14] The letter to U.S. Congress claims the gas tax increase "would disproportionately hurt lower income Americans already hurt by trying times in our economy." Both fail to note that poor and middle-class families will benefit most from public transit and other sustainable transportation options.

Although many organizations that promote the fossil fuel industry and reject the need to address climate change are secretive about their funding sources, a bit of digging often turns up oil, gas, and coal money, often from the Kochs. And most of their claims are easily debunked. In the case of the U.S. Heartland Institute, arguments stray into the absurd, like comparing climate researchers and those who accept the science to terrorists and murderers like the Unabomber and Charles Manson!

In some ways, it's understandable why fossil fuel advocates would reject clean energy, conservation, and sustainable transportation. Business people protect their interests—which isn't necessarily bad. But anything that encourages people to drive less and conserve energy cuts into the fossil fuel industry's massive profits. It's unfortunate that greed trumps the ethical need to reduce pollution, limit climate change, and conserve nonrenewable resources.

It's also poor economic strategy on a societal level. Besides contributing to pollution and global warming, fossil fuels are becoming increasingly difficult, dangerous, and expensive to exploit as easily accessible sources are depleted—and markets are volatile, as we've recently seen with precipitous oil price drops affecting the economies of oil-producing jurisdictions worldwide. It's crazy to go on wastefully burning these precious resources when they can be used more wisely, and when we have better options. Clean energy technology, transit improvements, and conservation also create more jobs and economic activity and contribute to greater well-being and a more stable economy than fossil fuel industries.

To reduce pollution and address global warming, we must do everything we can, from conserving energy to shifting to cleaner energy sources. Improving transportation and transit infrastructure is one of the easiest ways to do so while providing more options for people to get around.

If you live in an area with good public transit, make use of it. If not, lobby for improvements and stand up to those who would get in the way of better, cleaner transportation options. Those who profit from our continued reliance on fossil fuels will do what they can to convince us to stay on their expensive, destructive road. It's up to all of us to help change course.

### JOIN A CAR SHARE OR BUY A BETTER CAR

IN TODAY'S WORLD, it's not always possible to get by without a car or truck. They're often needed for work or to transport goods or people, it's sometimes necessary to get to places that aren't easily accessible by public transit, bad weather can get in the way of options like cycling or walking, and people often have to travel long distances that make other options impractical.

In many urban areas, car-share programs provide a good alternative. When you factor in the costs of purchase and depreciation, fuel, maintenance, insurance, and parking, car-share programs are much more cost efficient, even if you use them a lot. Many car-share programs use fuel-efficient cars, such as Smart cars, hybrids, and electric vehicles, and some offer options such as vans or cars with bike and storage racks. And, as their popularity increases, the vehicles become more readily available in an increasing number of urban centres and locations in those centres. They also reduce the overall space required for parking facilities and make people less reliant on cars, which reduces pollution and greenhouse gas emissions.

A 2015 study by the University of California, Berkeley's Transportation Sustainability Research Center concluded that one car-share vehicle can replace nine to thirteen vehicles and reduce greenhouse gas emissions per year per household by 34 to 41 percent, or 0.6 to 0.8 metric tons.[15] The study also found that U.S. households that switched from private car ownership to car-sharing saved $154 to $435 a month, and that people who used car-sharing also walked, bicycled, and carpooled more often.

The range of services is wide, from peer-to-peer to business-to-consumer to nonprofit co-ops. Some charge membership fees, whereas others are pay per use. Most offer mobile apps that allow users to locate and reserve vehicles. Even some traditional rental companies are getting into the car-sharing business, and employers are starting to offer car-sharing incentives or options to staff. The number of users worldwide has been growing rapidly, from fewer than 500,000 in 2006 to almost 5 million in 2014, according to Statista.[16] That number will grow even faster as more people realize the benefits. Factoring in other options such as ride-sharing transportation services like Uber and emerging driverless car technologies, ABI Research predicts car-sharing could reach 650 million users by 2030.[17]

Carpooling is another option, and can include informal or organized arrangements. Many employers encourage carpooling through facilitation or incentives such as free parking, and many jurisdictions provide incentives such as high-occupancy vehicle lanes. For those who only rarely need a car, even taxis and car rentals are often less expensive than owning.

For those who must own a car or truck, finding a fuel-efficient one is also good for the environment and the bottom line.

Many countries, including Canada and the U.S., have enacted fuel-efficiency standards for new vehicles, which is a start. But despite regulations passed in the U.S. requiring cars and light trucks to increase fuel efficiency to the equivalent of fifty-four miles to the gallon (4.4 litres per 100 kilometres) by 2025, research by the University of Washington and the Massachusetts Institute of Technology found that automakers have to get more serious about the way they make vehicles, in part because although they have improved engine efficiency, they have "also focused on improving acceleration times, adding new features and making other tradeoffs that cause cars to guzzle more gas."[18] And more cars on the road can offset emissions reductions from fuel efficiency.

Electric and hybrid electric vehicles also offer the promise of reducing emissions, though they aren't a perfect solution. Hybrid cars still use fossil fuels, and electric car technology doesn't resolve all of the issues, especially as the electricity still must come from somewhere, and in some places, that means coal-fired power plants. Car manufacturing is also energy intensive. And even fuel-efficient or electric cars require extensive infrastructure, from roads to parking lots, and take up huge amounts of space compared to options such as public transit and bicycles. Still, they're an improvement over inefficient vehicles that use only fossil fuels. Innovation and new technologies continue to make energy-efficient vehicles increasingly viable.

An Alberta company has even developed an electric car made out of hemp fibre.[19] Beyond reductions in fossil fuel use to power the car, the materials used to manufacture it are also more environmentally friendly. Hemp grows easily outdoors with little water or pesticides, and it can be used in lightweight but durable composites to build the cars.

One invention that partly avoids the problem of charging electric car batteries using electricity sources that may contribute to greenhouse gas emissions is U.S. inventor Charles Greenwood's inexpensive HumanCar.[20] It can operate as an exercise-based, human-powered vehicle or a plug-in hybrid electric. One to four people "row" the car to generate power. It can reach speeds of more than ninety-five kilometres an hour. Of course, it has its drawbacks, especially as one must be pretty healthy to operate it.

Cars powered by solar cells and hydrogen are also being developed, along with cars that use alternatives to fossil fuels, such as ethanol or biodiesel. Driverless cars are seen as another breakthrough in reducing emissions from vehicles. They can operate as on-demand taxi systems, reducing the need for massive private auto ownership, and they can anticipate road conditions, optimize speed and routes, and facilitate multimodal transportation. Most will be smaller, lighter, and powered by electricity.

Regardless of the kind of vehicle they use, drivers can also reduce their current gas consumption by as much as 20 percent with a few eco-driving tips—something the David Suzuki Foundation's Quebec office learned with its Drive Smart, or Roulez mieux, campaign. And adopting better driving habits demonstrates that doing what's right for the environment also makes good economic sense. Beyond saving money on gas, drivers can reduce wear on their cars, saving on maintenance and car replacement costs. One of the first things to do is make your transportation more efficient through planning. Instead of making separate trips to get to work and the store, combine the journeys.

Keeping a vehicle properly maintained, with regular tune-ups, including air filter and oil changes, and tires in good shape and properly inflated will allow you to go farther on less gas. Avoiding rush hour if possible and driving defensively can help ensure that the fuel you burn will get you to your destination more quickly and efficiently. Shutting off the engine if your car is stopped for more than a minute makes sense, too. Slowing down also helps. Going over the speed limit won't get you to your destination much faster, but it will burn more fuel. Other good habits include keeping your trunk clean—as less weight requires less fuel to transport—and using the car's accessories, especially air conditioning, sparingly.

As fossil fuels become scarce, and as our knowledge of the impacts of pollution and global warming increases, the benefits of doing all we can to use less gas just keep adding up. Savings at the gas pump will even offset the higher costs of the new fuel-efficient vehicles. The standards will also lead to more jobs, as new technologies are developed.

The need for solutions is obvious. Cars not only contribute to air pollution and greenhouse gas emissions, but they also cause water pollution from fuel-storage leaks, improper disposal of oil, and runoff from roads that washes into rivers, lakes, and oceans. Noise pollution, death and injuries from road accidents, and the impact of cars on the shape of the urban environment are all issues as well.

Technological developments are welcome, but maybe it's time we started rethinking our car culture as a whole. Can cars really be considered good forms of transportation when most in North America carry only the driver, and so use more than 900 kilograms of metal to transport less than 90 kilograms of human? Using a life-cycle analysis, which takes into account manufacture and disposal, as well as operation, we find that cars are inefficient products.

We aren't likely to do away with private cars in the near future, especially in rural areas with low population density. But we can at least start to think differently about our "need" for them.

Even in places like China, where car use is rapidly increasing, it's not all bad news. Although car culture is growing, the use of electric bikes is exploding. In 2008, people in China bought 21 million electric bikes, compared to 9.4 million cars. China now has 120 million e-bikes on the road, up from about 50,000 a decade ago.

We take our cars for granted, but really, they haven't been a part of our human culture for that long, and they needn't be an essential part forever.

## Be Energy Efficient

CONSERVING ENERGY IS one of the best ways to reduce your impact on the climate, especially in North America, where people tend to waste a lot. It's also a great way to save money! The average household in Canada and the U.S. uses about twenty times the electricity of a Nigerian household. Of course, that may be partly because of heating needs, but Canadian and U.S. households also use two or three times more electricity than a typical European household and about three and a half times the global average for households with electricity, according to the World Energy Council, or WEC.[21] The Natural Resources Defense Council, or NRDC, reports that up to 23 percent of electricity in U.S. homes "vanishes as 'standby' or 'always-on vampire power' feeding perpetually plugged-in electronics and appliances," even when they're not being used.[22] Cutting down on the use of energy-sucking electric goods makes a difference, as does unplugging devices and appliances when they are not in use. Many items, including cell phone chargers, continue to draw power if they are left plugged in, even when they are turned off or when no device is connected to them. The U.S. Department of Energy estimates that these "energy vampires" can add as much as 10 percent to your energy bill.[23] Using power strips or switches that allow you to turn off outlets in your home or workplace helps reduce energy by shutting off power when it is not needed. Power-management devices that put computers and other electronics into sleep mode or products that power off after a period of inactivity rather than switching to energy-sucking screensavers or slideshows is another way to conserve energy. Switching off lights, computers, televisions, and other devices when you're not using them is a no-brainer. What's next?

Energy-efficient lighting is improving all the time. Prices for light-emitting diodes, or LEDS, are coming down as efficiency improves. Many power utilities even offer rebates and other incentives for people who switch. Because it lasts much longer and uses up to 85 percent less energy to deliver the same amount of light as an incandescent bulb, one LED can save more than $100 over its lifetime. They now come in many sizes, shapes, and light levels and can fit in many existing outlets. To save even more on lighting, try doing more work during daylight hours, if possible.

When buying new appliances, look for energy-efficient ones, such as those with the Energy Star label. According to the NRDC, Energy Star washers and dryers will use 20 percent less energy over their lifetime, saving both energy and money. Wash clothes in cold or warm water rather than hot, and always rinse in cold. Dryers are energy hogs, using more energy than any other large appliance in most homes, so hang dry when you can. The NRDC says they represent 2 percent of total electricity consumption in the U.S. If you must use one, make sure the lint trap is kept clean, and dry heavy and light items separately, using the "normal" or "moisture sensor" setting.

Install a programmable thermostat that lowers home temperatures at night or when you're not home. Properly insulating your home makes a huge difference as well. Make sure to seal cracks, leaks, and spaces around doors and windows to ensure cold air doesn't enter in winter and warm air doesn't enter in summer. Double- and triple-glazed windows, patio doors, and skylights; good quality insulation; and a good heat-recovery ventilation unit will all keep your home more comfortable while saving money. A home energy audit may also be cheaper than you think and could end up helping you save money in the long run.

If you're lucky enough to be considering a new home, look into the many new technologies for making homes energy efficient. Net zero homes are designed to produce more energy than they consume, but they aren't without controversy. Some argue that they work best for large single-family dwellings on big lots—which isn't necessarily the most efficient form of housing for growing populations.

Homeowners can also install solar panels on roofs that get sufficient sunlight. Many power utilities also offer programs whereby customers pay a premium to support investment in and sourcing of renewable energy. According to the U.S. Department of Energy, "Nearly 850 utilities, including investor-owned, municipal utilities, and cooperatives, offer a green pricing option" in the U.S.[24] In some jurisdictions, buying power from electricity providers that use renewable energy is also an option. In the U.S., "competitive markets grew to 16.2 million MWh in 2014, a 12% increase from 2012." Renewable energy certificates, which allow consumers and businesses to buy credits based on the amount of renewable energy fed into the grid, are an option even in areas where green power is not available through the local utility.

Many jurisdictions and electricity providers offer incentives and programs to help people make homes and workplaces energy efficient. Among the strategies used by municipalities to finance infrastructure improvements that benefit homeowners are local improvement charges. The charges are repaid through property tax bills. With a property assessed payments for energy retrofits, or PAPER, program, homeowners can obtain financing for renovations to improve the energy efficiency of their homes and repay through a temporary fee on their property tax bill. The idea is to have cost savings from energy efficiency exceed payments.

Many homeowners need upfront financing to increase their homes' energy efficiency, but if they move before the financing is repaid, they are less likely to borrow. Having energy retrofit financing that remains with the property and doesn't add to owners' personal obligations allows homeowners to be responsible stewards of their homes, their wallets, and the environment. An added bonus is that these programs are designed to be delivered at no cost to municipalities and can benefit higher-level government budgets as well.

## Reduce Consumption and Waste

WASTING FOOD, THROWING away goods that might still be useful, and purchasing over-packaged products all create harmful environmental impacts in numerous ways. The effects on climate are felt at every step, from energy required to extract, produce, and transport goods and materials, to that required for their use, and emission of greenhouse gases such as methane from landfills when they're discarded. The U.S. Environmental Protection Agency estimates that 42 percent of the country's "greenhouse gas emissions are associated with the energy used to produce, process, transport, and dispose of the food we eat and the goods we use." Waste management accounts for 1 to 5 percent of the total.[25] Fossil fuels are often used in the extraction process—whether for food, trees, or minerals. Recycling normally doesn't require as much energy as extracting and processing raw materials, so it's a good way to ensure materials are reused rather than wasted and to reduce greenhouse gas emissions. Of course, the best way to cut down on waste, pollution, and emissions is to reduce the number of products you buy. Remember which comes first in the environmental slogan "Reduce, Reuse, Recycle." The next step is to make sure that you buy products that are made to last, and repair, share, or reuse products when you can.

It's important to question, as well, whether owning more "stuff" actually makes us happier. Most studies conclude that it doesn't. Often, people spend money on goods and services that they think will increase their status, or make their lives fuller. Sometimes it's just a way to compensate for a lack of fulfillment in life, to overcome loneliness, poor self-esteem, and disconnection from nature. People who focus on spending time with friends and family, spend time in nature, and learn to live with less are often happier and more fulfilled than those who devote their time, energy, and money to materialistic pursuits. Consumerism puts us on a collision course with enduring drudgery, the feeling that we never have enough, and growing environmental consequences.

For the true necessities in life, buying locally and sustainably produced goods helps cut down on impacts, such as emissions from transporting products over long distances. Planning your shopping trips to cut down on the number of times you might drive to the store also helps.

As noted in the previous section, reducing the amount of energy required to operate consumer products is also important.

Putting less in the trash bin is essential. People discard a lot of stuff when they no longer have a use for it. Landfills are packed with goods that could be repaired, donated, reused, or recycled. Many cities have electronic waste recycling programs or depots, and some retail outlets will take back electronic products such as computers and cell phones for reuse or recycling. Because some people get rid of computers and other items only to upgrade to something better, many of these products are thrown out even though they still work. Many charitable organizations and other institutions will take these products for use by people who couldn't otherwise afford them.

Many areas now have recycling programs, which help reduce waste and greenhouse gas emissions. Composting programs are also becoming more popular. These ensure that food waste is used to fertilize soils and grow more food, or converted to energy through burning. Even in areas without compost programs, people can get home composters, so that food waste can be turned into nutrients for the home or community garden.

Anything that ends up in the landfill, especially food, creates problems. Organic materials produce methane gas when they decompose. Methane is a greenhouse gas up to twenty times more potent than $CO_2$, so landfills can be a major source of greenhouse gas pollution. Although the best solution is to keep as much stuff as possible from going to the landfill in the first place, many jurisdictions are using the gas to generate electricity or provide fuel for boilers and other equipment.

Food waste is a particular area of concern, and one in which consumers can make a big difference. Every year a staggering one-third—1.3 billion metric tons—of the world's food is wasted after it has been harvested: 45 percent of fruit and vegetables, 35 percent of fish and seafood, 30 percent of cereals, 20 percent of dairy products, and 20 percent of meat. A 2015 study found Americans waste close to $200 billion on uneaten food, and Canadians throw away $31 billion.[26] In Canada, food waste cost estimates increased from $27 billion to $31 billion between 2010 and 2014.

These figures only account for 29 percent of the full cost of waste. They don't include factors such as labour, fuel to transport goods to global markets, inefficiency losses from feed choices used to produce meat and fish, or food left unharvested. As methodologies are improved and accounting becomes more inclusive, we're likely to find even higher waste figures. Dozens of studies across many countries with different methodologies not only confirm the increase in food waste but also suggest food waste is even higher and on the rise.

In a world where one in nine people doesn't get enough to eat—many of them children—this is unconscionable. According to the UN World Food Programme, poor nutrition kills 3.1 million children under the age of five every year.[27] It's the cause of almost half of child deaths in that age range. When it comes to feeding the world, distribution and waste appear to be greater problems than population. Yet we continue to destroy more forests, drain more wetlands, and deplete the oceans of fish to meet the needs of a growing world population.

Not only that, but food waste represents monumental economic losses—money that could be used to fund much-needed social and environmental programs. Money lost in North America would cover most of Canada's federal budget. Food waste in my home city of Metro Vancouver adds about $700 a year to a household's grocery bill.

Every morsel of food wasted represents unnecessary greenhouse gas emissions, conversion of natural ecosystems to agricultural lands, and disruptions to marine food webs. Based on 2007 data, the UN estimates that the equivalent of 3.3 gigatonnes of $CO_2$ emissions globally can be attributed to food waste.[28] Canada's total emissions, in comparison, are about 0.7 gigatonnes. If food waste were a nation, it would be the world's third-largest emitter.

We need to tackle food waste at all levels, from international campaigns to individual consumption habits. In 2015, the UN agreed to an ambitious global goal of reducing food waste by 50 percent by 2030 as both an environmental and humanitarian imperative. In early 2015, Metro Vancouver joined the international effort Love Food Hate Waste to meet municipal waste goals and encourage individual behavioural change. A similar U.K. campaign led to a 21 percent cut in food waste over five years. Grocery stores in France and other countries are offering discounts for misshapen produce under an "ugly fruits and vegetables" campaign. Businesses are using audits to map out where food waste is affecting bottom lines.

Food waste is a crime against the planet and the life it supports. Reducing it not only addresses food insecurity, it benefits everyone. Whether you're vegan, vegetarian, carnivore, locavore, or pescetarian, planning for zero food waste by buying only what you will eat and eating all that you buy is a positive step. We must take the same approach with all the products we use, buying what we need and making sure we get as much use out of it as possible, and recycling or repurposing products when they've reached the end of their useful life. Demanding through our consumer choices and with our voices that companies reduce wasteful packaging and create products to last is also crucial.

Perhaps most important is to change our way of thinking to realize that more stuff doesn't necessarily improve our lives.

## Eat Less Meat to Reduce the Heat

WILL VEGANS SAVE the world? Reading comments posted under articles about climate change or watching the film *Cowspiracy* might make one conclude they're the only ones who can. *Cowspiracy* boldly claims that veganism is "the only way to sustainably and ethically live on this planet."[29] But, as with most issues, it's complicated.

That's not to say there isn't some truth in those claims. If more people were to give up, or at least cut down on, meat and animal products, especially in over-consuming Western societies, the benefits for the environment and climate would be substantial. Animal agriculture produces huge amounts of greenhouse gas emissions, consumes massive volumes of water, and causes a lot of pollution—and eating too much meat can be bad for your health.

But getting a handle on the extent of environmental harm, as well as the differences between various agricultural methods and types of livestock, and between farmed and wild meat, and then balancing it all with the benefits of animal consumption and agriculture make it a complicated subject.

Even estimates of how much animal agriculture contributes to greenhouse gas emissions range widely, from about 14 to more than 50 percent of total global emissions. Part of that depends on the factors included, such as the impacts of clearing forests to make way for agricultural land. Most reliable sources put it near 14 percent, though, which is significant. The estimates are complicated by the fact that some of those emissions would be produced without the animals. Food is how we convert energy—solar energy—to fuel our bodies. Photosynthesis locks the sun's energy into carbohydrates in plant cells, like glucose, which plants can convert into proteins and lipids.[30]

Plants require less energy to produce equivalent amounts of nutrients—but in converting that energy, animals also produce useful nutrients that may not be as readily available from plant sources.

Agriculture contributes to climate change in a number of ways. Clearing carbon sinks such as forests to grow or raise food can result in net $CO_2$ emissions increases. Farming, especially on an industrial scale, also requires a lot of machinery that burns fossil fuel, as does processing and transporting agricultural products. And livestock emits large amounts of greenhouse gas emissions. Determining the overall contribution to climate-altering emissions is complicated by the fact that livestock agriculture accounts for about 9 percent of human-caused $CO_2$ emissions but far greater amounts of other greenhouse gases.

According to a UN Food and Agriculture Organization, or FAO, report, livestock farming produces 65 percent of human-related nitrous oxide, which is 296 times stronger than $CO_2$ in its global warming potential. It also contributes "37 percent of all human-induced methane (23 times as warming as $CO_2$), which is largely produced by the digestive system of ruminants, and 64 percent of ammonia, which contributes significantly to acid rain."[31] To complicate matters further, methane only stays in the atmosphere for about 12 years, and nitrous oxide for about 114, whereas $CO_2$ can remain in the atmosphere for thousands of years.[32]

Emissions also vary according to livestock. Pigs and poultry contribute about 10 percent of global agricultural emissions but provide three times as much meat as cattle—which are responsible for about 40 percent—and use far less feed.[33]

Plant agriculture can also contribute significantly to global warming. Rice cultivation, because of heavy nitrogen fertilizer use and methane created in rice fields, produces a lot of greenhouse gas emissions. And some types of agriculture—especially large-scale industrial farming—are more damaging than others. The amount of food that gets wasted, about one-third worldwide, also complicates matters.

Another factor is that people throughout the world have adapted genetically to different diets. Inuit have relied throughout their history on hunting—mostly whales, seals, and fish. Their environment is not amenable to growing large amounts of produce. According to a study in *Science*, "Inuit evolved unique genetic adaptations for metabolizing omega-3s and other fatty acids."[34] And in parts of the world where domestic cattle have not been common, people tend to be more lactose intolerant.

Despite the complexity, the bottom line is that cutting down on or eliminating meat and other animal products from our diets is necessary for protecting humanity from runaway climate change—and from many other environmental consequences, including water scarcity, degraded ecosystems, and pollution of waterways and oceans. The FAO reports that global demand for livestock products could increase by 70 percent by 2050 with an estimated human population of 9.6 billion if nothing is done to slow consumption.

Looking at worldwide statistics for meat consumption shows there's a lot of room to cut down in industrialized countries, where the average person consumed 95 kilograms of meat in 2015, compared to the 41-kilogram global average, and 31.6 in developing countries. People in South Asia eat less meat than anyone, at about 7.6 kilograms in 2015.[35]

A 2016 study by scientists at the U.K.'s Oxford Martin School found that global agriculture-related emissions could be cut by a third by 2050 if people followed health guidelines on meat consumption, and by 63 percent with widespread adoption of a vegetarian diet. With vegan diets, reductions were projected at 70 percent. The authors found that adoption of healthier diets with less meat and animal products could also reduce global health care costs by $1 billion a year by 2050.[36]

Although switching to better agricultural methods and encouraging local consumption could also reduce emissions, those are bigger topics. In the meantime, we can all do our part by at least cutting down on the amount of meat we eat, especially red meat, and by thanking those who have taken the significant step of adhering to a vegetarian or vegan diet.

Perhaps the best dietary advice for our own health and the planet's is from food writer Michael Pollan: "Eat food, not too much, mostly plants."[37]

## Divest from Destruction and Invest in the Future

IF PEOPLE KEEP digging up and burning fossil fuels, there's no hope of meeting climate change commitments made under the 2015 Paris Agreement. And if governments don't stick to their pledge to limit global warming to 1.5 or 2 degrees Celsius above preindustrial levels by 2050, we're in for some extremely hard times. To meet the goal, most experts agree that up to 80 percent of current oil, coal, and gas reserves must stay in the ground—and that makes fossil fuels a bad investment. Many analysts refer to them as stranded assets.

But that's not the only reason to avoid or pull investments from the fossil fuel industry. Profiting from industries that contribute to the endangerment of all humanity is simply unethical.

Avoiding investments related to the fossil fuel industry is difficult, though, because we live in a fossil-fuelled world. For individuals who put their money in mutual funds, for example, finding viable plans with no fossil fuel holdings can be a challenge, but it's certainly not impossible, especially as demand increases. Although individual investment choices, coupled with the growing public demand for fossil fuel–free choices, make a difference, the real momentum is at the institutional investment level. But even there, public pressure and campaigns have a huge impact.

Banks and investment advisers are increasingly cautioning investors against the dangers of keeping their money in industries related to fossil fuel. Many warn that climate agreements, government regulations, reduced demand, and market volatility put investors at risk. Some, such as HSBC, have suggested divestment as one route but note that investors may just want to pull their money from the most risky areas, such as coal and oil, or keep their investments so that they can engage with companies and have some influence over their decisions.

In rejecting a widespread call to withdraw about $70 million in fossil fuel money from its $1.3 billion endowment fund in 2016, the McGill University board of governors touted the latter as one reason. However, the university doesn't appear to see fossil fuels as a problem. In a report to the McGill board, the university's Committee to Advise on Matters of Social Responsibility wrote, "The Committee is persuaded that the beneficial impact of fossil fuel companies offsets or outweighs injurious impact at this time."[38] Many McGill alumni, including David Suzuki Foundation Quebec and Atlantic Canada director-general Karel Mayrand, have returned their degrees to the university in response.[39]

The failure of students, faculty, staff, and the public to convince boards at McGill, the University of British Columbia, Concordia University, Dalhousie University, and the University of Calgary to divest shows how entrenched the fossil fuel industry is when it comes to large investors. But it also shows that individuals play a major role in getting institutional money out of the industry. Although these campaigns didn't convince the boards to divest, they raised awareness around fossil fuel investments.

And the divestment campaign has had numerous successes. A growing number of universities, banks, pension funds, unions, churches, cities, insurance companies, individuals, and more have pulled funds, as has the *Guardian* newspaper, as part of its climate change campaign—all of which makes investments in the industry even shakier. The David Suzuki Foundation works with Genus Capital to ensure that none of its endowment fund is invested in funds related to fossil fuel, and to develop new strategies for ethical investing—which hasn't harmed financial returns. In fact, Genus reports, "fossil fuel free investing is proving more profitable than conventional investing."[40]

According to 350.org, the major force behind the divestment movement, as of late 2015, "more than 500 institutions representing over $3.4 trillion in assets have made some form of divestment commitment."[41] Divesting is just the start, though. As 350.org notes, reinvesting in "renewable energy, energy efficiency, and climate mitigation and adaptation infrastructure" not only helps the world shift away from fossil fuels but is also financially wise.[42] One option besides stock portfolios is to reinvest in initiatives that help the climate and bottom line, such as making buildings more energy efficient. According to the *Guardian* in 2015, "One conservative report calculated that investing $400,000 on efficiency in a 100,000 square-foot [9,300 square-metre] building would deliver $1.50 per square foot [$16 per square metre] in reduced energy costs over a similar building without the efficiency"—for an annual saving of $150,000![43]

Wastefully exploiting and burning fossil fuels is a thing of the past. There's no reason to put money into an industry that is destroying the systems that make human life possible, but there are many reasons to stop giving them money. It's time to invest in the future.

# Chapter 5

# AGRICULTURAL SOLUTIONS

**A**GRICULTURE PLAYS A major—and complicated—role in global warming. It's difficult to even determine the amount of emissions produced by crops and livestock. Most farms don't measure greenhouse gases, and emissions are offset to some extent because soils and plants absorb carbon. Researchers agree, though, that feeding the world's people has a tremendous impact on the climate.

The World Resources Institute, or WRI, estimates that 13 percent of global emissions come from farms, making farming the second-largest emitter, after the energy sector, including power generation and transport.[1] The U.S. Environmental Protection Agency puts it at 24 percent but includes deforestation in its calculations.[2]

Agriculture contributes to global warming in a number of ways. Farm machinery and infrastructure use fossil fuels that emit $CO_2$. But methane and nitrous oxide—which are far more potent than $CO_2$ but remain in the atmosphere for shorter times—make up about 65 percent of agricultural emissions. Much of the methane comes from cattle and the nitrous oxide from fertilizers and wastes. According to the WRI, "Smaller sources include manure management, rice cultivation, field burning of crop residues, and fuel use on farms." Net emissions are also created when forests or wetlands are cleared to make way for farms, as these carbon sinks usually absorb and store larger amounts of carbon than the farms that replace them. Transporting and processing agricultural products also contributes to global warming.

We all need to eat, though, so we can't just stop farming, especially with an expanding human population. Feeding more than 7 billion people with minimal environmental and climate impacts is no small feat. So, what's to be done? That obesity is epidemic in parts of the world while people starve elsewhere, and that an estimated one-third or more of the food produced gets wasted somewhere along the line, shows that improving distribution and reducing waste are good places to start—but although resolving some of those issues will address social justice and equity concerns, it will only go so far to curtail the agricultural contribution to climate change.

As mentioned in the previous chapter, reducing production and consumption of meat and animal products—especially beef—would be a major step in curtailing emissions, but even that won't get us all the way there.

The WRI points out that changes in crop-management and grazing-land practices can help, including better fertilizer management, conservation tillage, rotational grazing, altering forage composition, "and restoring degraded lands and cultivated organic soils into productivity."

Some suggest, however, that finding more efficient ways to feed as many as 9 billion people by 2050 means rethinking and retooling our agricultural systems. This is where it gets complicated. Soil loss and degradation, increased drought and flooding, and changing growing patterns caused by climate change add to the complexity.

It's clear that if we want to arrest global warming, along with a range of other environmental problems, we have to change the way we grow and produce food. Some argue the easiest way would be to adapt current systems by increasingly genetically modifying organisms to continually adapt to changing conditions—but this comes with numerous unintended consequences and is probably only a short-term solution. It also takes control from farmers and communities and increasingly puts it in the hands of large corporations.

Maps compiled by researchers at the University of Wisconsin-Madison show that 40 percent of the earth's land surface is now used for agriculture.[3] Given the massive geophysical alteration that entails, it makes sense to work as much as possible with nature to keep or restore to some extent balance in the earth's natural systems—not just in rural areas but in urban environments as well. Because environmental problems, and climate change in particular, are as much social justice as ecological issues, agroecology appears to be a better way to feed humanity than doubling down on industrial agriculture, from all angles: reducing pollution and chemical use, increasing biodiversity, protecting water, growing healthier food, and creating more equitable food systems. Encouraging people to cut down on or eliminate meat consumption also means that valuable farmland can be used to grow crops rather than raise livestock, which is more efficient from a nutritional perspective and is less damaging to the climate. Employing techniques like dark earth or terra preta for use in everything from home gardens to large farms will have immense climate benefits.

Increased urban agriculture makes cities more climate friendly, promotes food security and public health, and reminds us of our connection to food and nature.

We must realize that we are part of nature, and what we do to it we do to ourselves. The earth provides resources needed to feed humanity. We must learn to use those resources sustainably. In this chapter, we'll examine some of the debate around farming and its environmental impacts, especially those related to climate change. Are industrial agriculture and genetic modification the way to feed growing populations in a warming world, or should we be aiming to work more closely with natural systems? We'll also look at some novel ideas such as soil enhancement and urban agriculture.

## Are Industrial Agriculture and Genetic Modification the Answer?

OVER THE PAST half century, the world has moved increasingly to industrial agriculture—attempting to maximize efficiency through running massive, often inhumane livestock operations; turning huge swaths of land over to monocrops requiring liberal use of fertilizers, pesticides, and genetic modification; and relying on machinery that consumes fossil fuel and on underpaid migrant workers. Industrial agriculture has made it possible to produce large amounts of food fairly efficiently, but it also comes with numerous problems: increased greenhouse gas emissions; loss of forests and wetlands that prevent climate change by storing carbon; pollution from runoff and pesticides; antibiotic and pesticide resistance; reduced biodiversity; and soil degradation, erosion, and loss. Depletion of fertile soils is especially troubling, with losses estimated to be occurring up to one hundred times faster than they can regenerate with current industrial agriculture practices. Biodiversity loss refers to both a reduction in the number of crop varieties—more than 75 percent of plant genetic diversity has vanished over the past 100 years, according to the UN Food and Agriculture Organization[4]—and to reduced biodiversity among species that require diverse habitats for survival.

The "solution" many experts offer for feeding a growing human population is to double down on industrial agriculture and genetic modification. Some argue leaning more heavily on genetically modified crops, and perhaps even animals, is the only way to go. A new process called clustered regularly interspaced short palindromic repeats, or CRISPR, allows researchers to turn a specific gene on or off. It's being touted as a way to produce "plants that can withstand what an increasingly overheated nature has in store" and create "a more nutritious yield, from less plant," according to a 2015 Newsweek article.[5]

Those who oppose increasing reliance on genetic modification for agriculture are often accused of being "antiscience." Although it's true that some activists focus on potential health impacts of eating genetically modified organisms, or GMOs, and many studies have found no real evidence for such impacts, the technology comes with a host of other problems, some of them intertwined with industrial agriculture itself.

Many GMO proponents point to "golden rice" to illustrate the benefits of genetic modification and to criticize "counterproductive" attitudes of anti-GMO forces. The rice, which unlike many genetically modified products, is not patented by a large company like Monsanto, is modified to produce more vitamin A, thus potentially reducing infection, disease, and blindness among poor people who don't get enough of the vitamin. Noting that the International Rice Research Institute has itself admitted the rice hasn't yet proven to do much if anything to address the problem, Greenpeace Southeast Asia campaigner Wilhelmina Pelegrina told the *Washington Post*, "Corporations are overhyping 'Golden' Rice to pave the way for global approval of other more profitable genetically engineered crops. This costly experiment has failed to produce results for the last 20 years and diverted attention from methods that already work. Rather than invest in this overpriced public relations exercise, we need to address malnutrition through a more diverse diet, equitable access to food and eco-agriculture."[6]

A number of researchers agree. Washington University researcher Glenn Stone, initially a golden rice supporter, said, "The rice simply has not been successful in test plots of the rice breeding institutes in the Philippines, where the leading research is being done."[7]

Industrial agriculture and increased genetic modification ignore how natural systems function and interact and assume we can do better. History shows such hubris often leads to unexpected negative results. Excessive use of pesticides such as DDT is just one example of human innovation and "dominance" over nature that came back to bite us. People thought DDT was a benign wonder chemical that would reduce diseases spread by mosquitoes and protect crops from insects. Then, in 1962, biologist Rachel Carson published *Silent Spring*,[8] which showed that the chemicals biomagnify as they move up the food chain. In other words, higher concentrations of the chemicals accumulate in fat cells of animals throughout the food chain, with the highest concentrations found in top predators, including humans. Predatory birds, such as eagles, were hit especially hard by widespread DDT use. Of course, our use of fossil fuels, once thought to be an entirely beneficial fuel that would improve lives and give people more freedom and mobility, is another example of how the lack of a full understanding of natural systems can lead to dire consequences.

## Working with Nature through Agroecology

INDUSTRIAL AGRICULTURE IS also turning out to be an example of our dangerous lack of understanding of nature. For that reason, a growing number of researchers argue that working with nature would be far more efficient and less damaging. Agroecology combined with "participatory and evolutionary plant breeding" beats GMOs. According to agricultural scientist Salvatore Ceccarelli, writing in *Independent Science News* in 2016, "At best, GMOs can only be a short-term solution to any particular problem, but in every case they have created an often more serious problem (resistant weeds, insects or disease) that requires a new GMO and/or more chemical use. They also make a farmer completely dependent on the company producing the GMOs and chemicals."[9]

Genetically modifying plant crops, and even some livestock and fish, may appear to be a way to create food crops and animals that can withstand rapidly changing weather and climate patterns, and to increase yields, but, as Ceccarelli notes, it shares the same problems with all industrial agriculture methods, and probably comes with many unexpected long-term consequences.

By participatory and evolutionary plant breeding, Ceccarelli means the practice of having individual farmers collect, save, and plant a variety of seeds, using subsequent seeds from crops that fare best to plant future crops. In this way, crops adapt to changing soil, pest, and climate conditions, and biodiversity is maintained while farmers retain control over the seeds and crops. "Combining seed saving with evolution and returning control of seed production to the hands of farmers, it [evolutionary plant breeding] can produce better and more diversified varieties," he wrote. "These can help millions of farmers to reduce their dependence on external inputs and their vulnerability to disease, insects and climate change and ultimately contribute to food security and food safety for all."[10]

In a 2016 *Huffington Post* article, sustainability author Jeremy Lent describes agroecology, which is used in Latin America and gaining acceptance in the U.S. and Europe, as any method that "designs and manages food systems to be sustainable, enhancing soil fertility, recycling nutrients, and increasing energy and water efficiency."[11] The goal, writes University of California, Berkeley, agroecology professor Miguel Altieri, "is to design an agroecosystem that mimics the structure and function of local natural ecosystems; that is, a system with high species diversity and a biologically active soil, one that promotes natural pest control, nutrient recycling and high soil cover to prevent resource losses."[12]

Lent points to a study by the Rodale Institute that concluded global adoption of agroecological practices such as using "cover crops, compost, crop rotation and reduced tillage" could "sequester more carbon than is currently emitted."[13]

Agroecology is not just theory. According to former UN special rapporteur on the right to food Olivier De Schutter, "Today's scientific evidence demonstrates that agroecological methods outperform the use of chemical fertilizers in boosting food production where the hungry live—especially in unfavourable environments."[14] He further notes that, "to date, agroecological projects have shown an average crop yield increase of 80% in 57 developing countries, with an average increase of 116% for all African projects."[15]

## Dark Earth Could Herald a Brighter Future

ONE ESPECIALLY PROMISING agroecological technique has emerged with renewed interest in a soil-building method from the distant past called "dark earth," or "terra preta," which involves mixing biochar with organic materials to create humus-rich soil that stores large amounts of carbon. In the book *Terra Preta: How the World's Most Fertile Soil Can Help Reverse Climate Change and Reduce World Hunger*,[16] Ute Scheub and coauthors claim increasing the humus content of soils worldwide by 10 percent within the next fifty years could reduce atmospheric $CO_2$ concentrations to preindustrial levels. However, some critics claim such figures, and biochar's climate benefits overall, are exaggerated, especially as we don't know how long the carbon will remain sequestered in the soils and how much biomass will need to be converted to biochar to realize the intended benefits—or what environmental impacts that will have.[17] Some are also wary of mitigating climate change with geoengineering, a category biochar falls into, fearing that spending money and resources on these schemes might shift resources away from methods to reduce fossil fuel consumption.

But dark earth undoubtedly stores carbon, which is just one of its many exciting possibilities. It also enhances and activates soils so that they produce higher yields, helps retain water, neutralizes toxins, produces fewer methane and nitrous oxide emissions, and prevents erosion. It's more alive with bio-diverse microorganisms, making it easier for crops to adapt to changing conditions. (*Terra Preta*'s authors point out that just a handful of healthy soil contains billions of microscopic organisms, whereas many modern agricultural techniques leave the soil depleted and dead, requiring greater use of fertilizers to grow crops.[18]) Dark earth is also a good way to recycle nutrient-rich food scraps, plant wastes, and even human and animal urine and feces, rather than allowing them to pollute soil, water, and air through burning and runoff. It can be made with almost any kind of organic waste, and may be a viable way to deal with growing volumes of sewage.

Biochar is a form of charcoal made via pyrolysis—heating organic wastes in a low-oxygen environment. According to Scheub, "If you pyrolyze organic wastes, up to 50 percent of the carbon, which plants have extracted from the atmosphere in the form of carbon dioxide, is converted into highly stable carbon, which can persist in soils for thousands of years."[19] Some biochar systems can also produce heat and electricity from burning the gases that accrue. As well as carbon, bio-char retains nutrients such as nitrogen and phosphorus, and because it's porous, adding it to soils and compost helps them store nutrients and water.[20] The soils also emit less of the greenhouse gases methane and nitrous oxide than other soils.

Western scientists first studied terra preta in 1874, when Canadian-born Cornell University professor Charles Hartt and his team found patches of dark, fertile soils, several yards deep, along parts of South America's Amazon River, where earth is normally low in nutrients and organic matter.[21] Later archaeological research found evidence, including pottery shards, that indigenous communities created the soils over many years beginning more than five thousand years ago.

Scientists have since shed more light on the technique. Because the ancient practice is still employed in Liberia and Ghana, Africa, scientists from Sussex, Cornell, and other universities were able to compare dark earth to soils nearby where the technique isn't used.[22] They found dark earth contained 200 to 300 percent more organic carbon and could support "far more intensive farming."

Cornell University lead author Dawit Solomon was surprised that "isolated indigenous communities living far apart in distance and time" achieved similar results unknown to modern agriculturalists. "This valuable strategy to improve soil fertility while also contributing to climate-change mitigation and adaptation in Africa could become an important component of the global climate-smart agricultural management strategy to achieve food security," he said.[23]

There are potential drawbacks and barriers to wide-scale employment of dark earth, but most can be overcome.

The main concern among environmentalists and others is that it could lead to deforestation as trees are cut to provide biomass to make biochar, or that land will be turned over to produce biomass rather than food crops. Others have warned that it could also lead to genetic manipulation of trees for biomass. Scheub argues this can be overcome with biochar certification, which is now available in Europe and North America. And, she notes, "Although the potential for abuse is always a possibility, there is... an enormous amount of underused organic waste freely available for carbonization, making the felling of forests for biochar production highly unlikely and economically absurd."[24]

Because plants draw nutrients from the soil and release substances that can accumulate, terra preta is subject to the same problems as other forms of agriculture if monoculture—planting the same crops in the same place year after year—is practised, but crop rotation and mixed cultivation can overcome this.

As for the main barrier, *Terra Preta's* authors write, "Similar to decentralized renewable energy, which in the long term threatens the existence of the central energy corporations, the increasing spread of terra preta methods would ultimately cut into the profits of agro-industrial players."[25] Reduced reliance on chemical pesticides, fertilizers, and patented seeds, as well as machinery needed to till and plow, means more control over—and more of the profit from—farming is taken from the agricultural industry and put back into the hands of farmers. We can surely expect resistance from industry to anything that threatens its profitability.

Scheub and her coauthors say the technique can be used on any scale, from home and community gardens to large farms. *Terra Preta* includes instructions for creating biochar and enhanced soils.

Dark earth won't solve all our climate problems, but combined with reducing fossil fuel use, it could make a huge difference while addressing many agriculture, food security, and hunger issues. As Scheub argues, "Sequestering carbon through farming is not only the most expedient and the most ecological way to reduce $CO_2$ levels in the atmosphere, it is also the most cost efficient."[26]

## How Much Food Can Cities Produce?

HUMANS ARE RAPIDLY becoming city dwellers. According to the United Nations, "The urban population of the world has grown rapidly from 746 million in 1950 to 3.9 billion in 2014," and 66 percent of us will likely live in urban environments by 2050. The number of megacities is also skyrocketing, from ten in 1990 to twenty-eight in 2014—home to more than 453 million people. That's expected to grow to forty-one by 2030.[27]

That's spurred a renewed and growing interest in producing food where people live. Urban agriculture could help take pressure off rural land bases while providing numerous other advantages. This growing trend won't resolve all our problems around food production and distribution, but it offers many benefits. From balcony, backyard, rooftop, indoor, and community gardens to beehives and backyard chicken coops to larger urban farms and farmers' markets, growing and distributing local food in or near cities brings fresh, nutritious local food to the table.

But it does so much more. As writer and former Vancouver city councillor Peter Ladner (also the David Suzuki Foundation board chair) writes in *The Urban Food Revolution: Changing the Way We Feed Cities*, "When urban agriculture flourishes, our children are healthier and smarter about what they eat, fewer people are hungry, more local jobs are created, local economies are stronger, our neighborhoods are greener and safer, and our communities are more inclusive."[28] Local and urban agriculture can also help reduce greenhouse gas emissions that fuel climate change, and it can recycle nutrient-rich food scraps, plant debris, and other "wastes." And because lawns require a lot of water, energy for upkeep, and often pesticides and fertilizers for little more than aesthetic value, converting them to gardens is environmentally beneficial and provides food sources.

A 2016 study out of the U.S. Johns Hopkins Center for a Livable Future at the Bloomberg School of Public Health found that urban agriculture could "increase social capital, community well-being, and civic engagement with the food system,"[29] as well as "supplementing food security on various scales— providing ecosystem services to cities, improving the health of the residents, and offering opportunities to build skills and generate additional income."[30] Gardening has also been shown to have therapeutic benefits.

Although the study found many climate benefits—including reduced emissions from transporting food, carbon sequestration by vegetation and crops, possible reduced energy and resource inputs and waste outputs, and enhanced public interest in protecting green spaces—it noted some limitations. These include possible increases in greenhouse gas emissions and water use "if plants are grown in energy- or resource-intensive locations," less efficiency than conventional agriculture in terms of resource use and transportation emissions, and reduced population density leading to more emissions related to driving.[31] It also noted that, depending on practices, it can cause pollution from pesticide and fertilizer use. Overall, though, the study found urban agriculture to be a positive trend but concluded, "Many of the demonstrated benefits of urban agriculture efforts will only be achieved with adequate local, state, and federal governments' long-term commitment of support."[32]

Urban agriculture itself isn't new. During the First and Second World Wars, Canada, the U.S., the U.K., Australia, and

Germany all encouraged "victory gardens" to aid the war effort by reducing pressure on food systems. Gardens and chicken coops sprang up on private and public lands, in yards, parks, school fields, golf courses, railway edges, and vacant lots. Instead of mowing sports fields, sheep were put on them to graze and keep grass in check. During the First World War, 5 million urban or near-urban gardens were planted in the U.S., producing food valued at $1.2 billion by the end of the war. Peter Ladner notes that, during the Second World War, the U.K. had 1.5 million allotment plots producing 10 percent of the country's food, including half its fruit and vegetables, and the U.S. supplied 40 percent of its domestically consumed produce with more than 20 million home gardens by war's end.[33]

Granted, there were fewer people and more open spaces then, but it's still possible to grow a lot of food in urban areas, especially with techniques involving composting and creating soils such as terra preta, or dark earth. Ladner writes that Toronto plans to supply 25 percent of its fruit and vegetable production within city limits by 2025, and that a study from Michigan State University concluded Detroit could grow 70 percent of its vegetables and 40 percent of its fruit on 570 vacant lots covering two thousand hectares of city land.

The David Suzuki Foundation's Homegrown National Park project in Toronto has demonstrated the growing public interest in urban gardening and green space restoration. What started in 2013 as "an effort to create a butterfly-friendly corridor through the City of Toronto, along the former Garrison Creek" has become a growing movement of people who have made planters from old canoes, built rain gardens, initiated a number of popular public events, and planted more than fifteen thousand pollinator-friendly native plants.[34]

Cities needn't be wastelands of car-choked roads and pavement. Incorporating food production into ever-expanding urban areas makes them more livable and enhances the natural systems that keep us alive and healthy.

Chapter 6

# TECHNOLOGICAL SOLUTIONS

A BUNDANT, CHEAP FOSSIL fuels have driven explosive technological, industrial, and economic expansion for more than a century. The pervasive infrastructure developed to accommodate this growth makes it difficult to contemplate rapidly shifting away from coal, oil, and gas, which creates a psychological barrier to rational discourse on energy issues.

The ecological and true economic costs of energy use force us to scrutinize our way of living. And because our infrastructure doesn't allow us to entirely avoid fossil fuels, we must face the contradiction between how we should live and constraints against doing so.

Sustainability requires conserving energy and producing what we require with minimal ecological upset. Yet the inability to consider the need to shift quickly from fossil fuels means governments and industry look to mega-technologies such as carbon capture and storage, or CCS, to justify inaction on reducing greenhouse gas emissions, while dismissing solar and wind as impractical, too expensive, or unable to meet energy needs. Nuclear power may be an alternative to fossil fuels, but it's extremely expensive and would not be online were it not for enormous subsidies. Nuclear fuel is also finite, so costs will rise while the problem of radioactive-waste disposal remains unsolved. These large-scale solutions also tend to concentrate power in the hands of governments and wealthy corporations, as they require massive investment, central planning, and, often, intensive security measures.

You've probably heard many arguments against shifting to clean energy: wind doesn't always blow, sun doesn't always shine, the technology's not advanced enough, installations take up too much space, renewable energy is too intermittent and difficult to integrate into a system that relies on baseload (power that always runs) that can only come from fossil fuels, nuclear power, or large-scale hydro. And so we carry on, rushing to squeeze every last drop of oil and gas from the ground, using increasingly difficult and destructive methods such as fracking, deep-sea drilling, and oil sands extraction, with seemingly little concern for what we'll do after we've burned it all. But many of these issues are engineering problems rather than renewable energy issues. Surely if we can split atoms for energy we can find a way to deal with cloudy skies.

A lot of research, including some by the David Suzuki Foundation, working with the Trottier Energy Futures Project, is also challenging those skeptical assumptions. "Canada has vast renewable energy resources in the form of hydropower, solar, wind energy, and biomass, as well as geothermal, wave, and tidal resources that are many times larger than current or projected levels of total fuel and electricity consumption," a 2013 Trottier report, *An Inventory of Low-Carbon Energy for Canada*, concluded.[1]

Those findings are confirmed by research and experience elsewhere in the world. A 2013 study by engineers at Stanford University reported, "It is technically and economically feasible to convert New York's all-purpose energy infrastructure to one powered by wind, water and sunlight," and doing so "shows the way to a sustainable, inexpensive and reliable energy supply that creates local jobs and saves the state billions of dollars in pollution-related costs."[2] A 2013 article in the *New York Times* pointed to research by the Paris-based International Energy Agency, showing "thirteen countries got more than 30 percent of their electricity from renewable energy in 2011."[3]

The Stanford study's lead author, engineering professor Mark Z. Jacobson, told the *New York Times*, "You could power America with renewables from a technical and economic standpoint. The biggest obstacles are social and political—what you need is the will to do it."[4] It would be even less of a challenge in a country like Canada, which produced more than 63 percent of its electricity with renewable sources in 2011 (much of it from hydro). The U.S. produced 12.3 percent.

The biggest obstacles in shifting to clean energy may be social and political, but one of the greatest challenges is creating a "smart" electricity grid. As former Trottier Project managing director Ralph Torrie said of Canada's system, we'll have to replace our antiquated grid with one that "will use information technologies to balance a wider range of supply sources, energy storage, interprovincial transfers of electricity, and a wide variety of energy management and efficiency tools."[5]

Other challenges include the costs and the impacts of renewable energy installations on ecosystems and wildlife. And with biofuels, the sustainability of source materials and effects on land and food supplies must also be considered. But these are far from insurmountable. Fossil fuel and nuclear power sources are also extremely costly and have far greater environmental impacts. And many studies show that moving to renewables creates jobs and contributes to economic health.

New technologies are being developed all the time, too— from those that take advantage of unconventional energy sources such as installing turbines in sewage and water pipes to devices that use energy from exercise equipment to generate power. Biomimicry, or replicating natural processes, is also showing some promise. One technology uses artificial leaves to convert solar energy to chemical energy by photosynthesis, as plants do.

Sure, changing the way we generate and supply energy is a challenge. But the alternative—to carry on polluting air, water, and soil, and putting our future at risk with global warming— isn't pretty. We've faced and overcome many challenges before. When people have mobilized resources in the past, we've been able to accomplish a lot in relatively little time—from defeating the fascist threat in the Second World War to putting people on the moon. Finding smarter ways to power our societies is something we can and must do. In this chapter, we'll look at some of the technologies that are helping the world shift from outdated, polluting, climate-disrupting fossil fuel power sources.

## Smart Grids and the Baseload Argument

ONE ARGUMENT USED against making a transition from fossil fuels to renewable energy is that because some renewable energy, such as wind power, is intermittent, we need baseload power sources such as large fossil fuel or nuclear power plants or massive hydro dams. But emerging and existing technologies may allow us to use renewables for baseload power, and many experts argue that we don't need baseload at all—that it's outdated. The website Skeptical Science notes that if baseload is required, technologies and sources such as concentrated solar thermal, enhanced geothermal, wind compressed air energy storage, and pumped heat energy storage can all play a role.[6]

With conservation and improved efficiency, though, along with better storage and smart grid management, we could switch to renewables without the need for large-scale baseload. Australian wind power researcher Mark Diesendorf goes as far as to argue that the main obstacle to renewable power development is the "operational inflexibility of base-load power stations."[7] He says the fossil fuel and nuclear sectors, as well as industries that depend on them, such as aluminum and cement manufacturers, promote the "baseload fallacy." (Nuclear plants must also be shut down for weeks at a time for refuelling.) As a 2015 article by Michael Mariotte on the GreenWorld blog noted, "That's because the kind of grid that works for the variable renewables—a fast, nimble grid where power from different sources scattered in different locations can be ramped up and down quickly depending on where it is being generated and where it is needed—doesn't work well for baseload plants, especially nuclear reactors, which cannot ramp up and down quickly."[8] Mariotte argued that when you add in new storage technologies, the idea of large, polluting, twenty-four-seven power plants is an "anachronism."

In a 2010 article for ABC Science, Australian energy expert David Mills wrote, "More fundamental to meeting our energy demands is the ability to match inflexible sources of power—those that can only generate energy at certain times such as wind—with flexible sources of power—those that can generate and store energy such as solar."[9] Miller and his colleagues conducted an analysis to demonstrate that the U.S. could have supplied 100 percent of its electricity in 2006 "on an hour-by-hour basis for the whole year solely from wind and solar energy. No baseload power required."

As writer David Roberts pointed out in a 2012 article on Grist, Germany decided that baseload and renewable energy technologies aren't compatible. Conventional power grids use baseload, medium load, and peak load sources, but Roberts wrote, "if you have enough renewables, they completely take over the space once occupied by baseload." To supply the demand, or residual load, that renewables can't cover, you need flexible and responsive options. And that will come from "a combination of demand-side measures (conservation, efficiency, and 'peak shaving' through demand response), energy storage, a much smarter grid, and dispatchable power sources."[10]

In the short term, Germany plans to use natural gas and imports as its "dispatchable" power source, but with emerging storage technologies, including converting renewable energy to synthetic natural gas or biogas, Germany could stop using all fossil fuels in its power sector.

Better grid management systems appear to be key to moving away from baseload. According to Steve Holliday, CEO of National Grid, which operates gas and power transmission networks in the U.K. and northeastern U.S., "The world is clearly moving towards much more distributed electricity production and towards microgrids. The *pace* of that development is uncertain. That depends on political decisions, regulatory incentives, consumer preferences, technological developments. But the direction is clear." He added, "The idea of baseload power is already outdated. I think you should look at this the other way around. From a consumer's point of view, baseload is what I am producing myself. The solar on my rooftop, my heat pump—that's the baseload." He also noted that storage is no longer an issue, as computer technologies allow systems "to ensure power is consumed when it's there and not when it's not there."[11]

Writing for *Vox* in 2015, David Roberts noted that grid rules and markets must be revised if we are to move away from baseload to more flexible renewables. "There are three changes, broadly speaking, that need to be accelerated in all electrical grids: resource flexibility, operational flexibility, and integration," he wrote. By "resources," Roberts means "anything grid operators can use to ensure that supply meets demand—not only power plants, but programs that manage demand as well." To accommodate fluctuations in variable renewable energy supply, those resources must be able to "more nimbly scale up and down." That means relying less on large coal and nuclear power plants. To encourage the shift requires central planning and "making the value of flexibility more visible in markets, to create incentives for investment in flexible resources." According to Roberts, that could mean removing price caps in energy markets, tweaking capacity markets to better value ancillary services (e.g., speed and responsiveness), or "creating other, parallel forward markets for balancing or time-shifting services."[12] Increased market competition will also help.

Roberts noted that a range of tools and technologies make demand management more practical, from systems that provide incentives for industrial users to reduce operations during peak demand times to getting homeowners to reduce water heater temperatures when grids are congested to making use of batteries in electric cars to store and supply power.

Operational flexibility means employing methods such as using new technologies for more agile scheduling and dispatch; using better weather forecasting tools to manage supply; expanding locational pricing to vary "the price of energy based on where it is produced, reflecting geographical differences in demand and grid congestion"; and tweaking markets "to more accurately value services beyond energy."[13]

Grid integration, which allows power from renewable energy to be drawn from a large area, Roberts noted, means "supply becomes steadier, fluctuations become less sharp, and prediction becomes more tractable."[14]

If we don't need baseload, why are we so slow in moving away from large polluting power plants that run constantly and require power to be transmitted over long distances through inefficient power lines? Mariotte suggested part of the reason may be that our economic thinking is as anachronistic as our ideas about energy supply: renewable energy is inexpensive, whereas large gas, coal, and nuclear plants are costly but continue to generate profits as well as electricity. Because twentieth-century baseload power is incompatible with most smart-grid technologies that distribute power from various sources according to supply and demand, development of new electricity technologies often means shutting down old-school baseload power stations, even if they have not reached the end of their useful life cycle. That means an end to the profits for the plants' owners and possibly even losses if the plants haven't been operating long enough to recoup construction and start-up costs. And because renewable energy is cheaper to produce, it's not as profitable to supply.

But outdated economic arguments aren't enough to justify continued use of outdated electricity systems. As Mariotte noted, "Why throw carbon dioxide into the air and tritium into the water and generate lethal radioactive waste just to keep dirty and usually more expensive power plants operating just for those few hours in the week when they might be useful? With storage, they're not needed, or even particularly useful, at all."[15] And when you factor in the external costs of pollution and greenhouse gas emissions, as well as other environmental damage related to fossil fuels and nuclear energy, those anachronistic energy sources aren't as cheap as they appear. It's another example of the absurdity of economic systems that rely on constant growth, consumption, and waste to generate profits.

Not only do we have to change our ways of thinking about energy use and distribution, we have to change our outdated ways of considering economics and progress. Computer technology is already forcing that change. Many things that were once expensive, such as long-distance communication and information sharing, are now easy and inexpensive. Continuing to obtain electricity from polluting sources that are putting all of humanity at risk by contributing to global warming is far too costly in every way imaginable. We're well into the twenty-first century; it's time to move away from twentieth-century technology.

## Solar: A Brilliant Way to Get Energy

EXCEPT FOR NUCLEAR and geothermal, virtually all energy we use comes from the sun in some form.[16] As sunlight reaches the earth, it powers heat transfers that move the air, supplying wind power. Tidal power comes in part from the sun's gravitational pull. The sun also evaporates water, contributing to the hydrologic cycle that fills reservoirs for hydropower.

Even fossil fuels—coal, oil, and gas—are forms of solar energy, created hundreds of millions of years ago, when plants absorbed and converted sunlight through photosynthesis, then retained that energy when they died, decayed, and became compacted and buried deep in the earth, along with the animals that ate them. Wood, peat, dung, and other plant-based fuels are a less concentrated form.

Most people think of solar energy in its direct form: energy harnessed in a variety of ways from the sun's rays as they hit Earth.[17] Technologies range from windows and water tanks placed strategically to make use of the sun's energy, to photovoltaic cells (or solar panels), to large mirrors that concentrate the sun's heat in order to boil water and drive turbines.

Unlike fossil fuels, or uranium-dependent nuclear power, the energy source is free, inexhaustible, and nonpolluting with no troublesome by-products like radioisotopes or carbon dioxide. It can be used for a variety of applications, from providing power for a single streetlight to generating electricity for a home to keeping satellites and factories going. But solar also has some disadvantages. Without energy storage systems, it only works when the sun is shining, and technologies to harness it can be costly. Solar installations that can provide power for large areas can also take up a lot of space, and some technologies rely on rare materials that must be mined, with environmental consequences.

Because of rapid technological advances, dropping prices, and the many advantages of solar power, it's become one of the fastest-growing sources of renewable energy worldwide, with installed capacity growing on average 43 percent a year since 2000, according to the World Economic Forum.[18] Still, as a 2015 MIT report points out, solar was only generating about 1 percent of global electricity in 2015.[19] New technologies for harnessing and storing the sun's energy could drive that percentage much higher.

About 90 percent of currently installed solar capacity uses crystalline silicon wafer–based photovoltaic cells. As a Phys. org article notes, they're nontoxic, abundant, and reliable, but the silicon wafers are thick and rigid and somewhat expensive to manufacture.[20] New technologies are increasing efficiency, with cells that are smaller, less rigid, and often more affordable, but they can come with other problems. Some use materials that "involve rare and/or toxic metals."

The website notes that promising "third generation" developments include "thin-film solar photovoltaic employing dye-sensitized, organic, quantum dot or perovskite solar cells and novel combinations of semiconductor materials, as well as concentrators." (Perovskite refers to the crystal structure of the materials that absorb light, based on the structure of the naturally occurring mineral of the same name. Often they are lead- or tin-based.)[21] Technology is also being developed to mimic photosynthesis, converting sunlight to electrons with nanotechnology and light-absorbing compounds and delivering the electrical energy "to customized catalysts that convert water and $CO_2$ into oxygen and chemical fuels." Artificial photosynthesis is also being studied as a way to capture and convert $CO_2$ emissions to generate fuels, plastics, drugs, and other products![22]

Efficiency ratings measure the amount of sunlight hitting a panel that will get turned into usable electricity. So far, the silicon cell is the most efficient at up to 26 percent, compared to 10 to 20 percent for many of the emerging technologies. But the new thinner solar cells can absorb the same amount of light with smaller panels. Silicon is also plentiful. The MIT researchers examined scenarios to determine how much photovoltaic material would be needed to satisfy 5, 50, or 100 percent of global electricity demand in 2050, and concluded, "Meeting 100 percent of 2050 global electricity demand with crystalline silicon solar cells would require the equivalent of just six years of current silicon production. Such a scale-up of production by 2050 is certainly feasible, so materials constraints are not a major issue for silicon."

The MIT researchers noted, "In contrast, using commercial thin-film technologies such as cadmium telluride to supply the bulk of projected electricity demand would require hundreds of years of producing key materials at current rates. The needed growth in annual production of those materials between now and 2050 would be well beyond the realm of historical precedent." Existing thin-film technologies also require many years to mine and process the materials, which, according to the MIT study, means they can be used, but will probably not be the dominant technologies. The newer emerging thin-film technologies show more promise, as many use "abundant primary metals that are produced in high volume." Quantum-dot cells would only require twenty-two days of minerals production, and perovskites would require about three years to produce the materials to meet 100 percent of global electricity demand in 2050.

Placement of solar cells is also raising interesting possibilities, including paving roads, parking lots, and bike lanes with durable panels that not only generate power but also melt snow and ice, and can display warning messages and traffic lines. Most projects are still in the test phase, but a seventy-metre Dutch solar bike path generated more energy than expected after installation in November 2014—enough to power a single-person home for a year, with expectations that it would generate seventy kilowatt hours per square metre per year.[23]

As mentioned in the previous section, experts such as National Grid CEO Steve Holliday predict solar technology could make the need for baseload power and even transmission grids obsolete.[24]

One of the more recent major developments in solar energy is taking place in Morocco. Phase I of a massive $9 billion solar power project near the city of Ouarzazate on the edge of the Sahara Desert has already been completed, with full operation slated for 2018. The facility—which will be the world's largest concentrated solar power plant, supplying electricity to more than a million people—will use 500,000 crescent-shaped mirrors arranged in eight hundred rows that shift to capture the sun's energy.

According to the *Guardian*, "Each parabolic mirror is 12 metres high and focussed on a steel pipeline carrying a 'heat transfer solution' (HTF) that is warmed to 393C as it snakes along the trough before coiling into a heat engine. There, it is mixed with water to create steam that turns energy-generating turbines."[25] The heat transfer solution is pumped to a heat tank with molten sands to store energy for times when the sun isn't shining.

The Moroccan government decided to shift to renewable energy because the country is not an oil producer and was importing and subsidizing fossil fuels. Its goal is to get half its electricity from renewables by 2020, using equal shares of solar, wind, and hydro.

Solar is currently viable and affordable enough (especially with tax credits in some jurisdictions, including the U.S.) that it can be easily installed on rooftops to generate electricity for homes, public institutions, and businesses. It works as well or better in cold regions, such as high altitudes, which are exposed to more sunlight, in part because many solar cells work more efficiently and are subject to less damage in cold temperatures.[26] Obviously, though, they are more useful in areas with abundant sunlight, even if temperatures are cold. It's becoming an increasingly important part of the energy mix, especially as energy transmission-grid and battery storage system technologies become more sophisticated and efficient.

Every hour, the sun bathes the earth with enough energy to supply our needs for more than a year. There's no reason we can't harness more of it to cut back on polluting, climate-altering fossil fuels.

## The Answer Is Blowing in the Wind

ENERGY FROM WIND is one of the fastest-growing developments in the shift from fossil fuels, and one of the most controversial. Although wind only supplied about 3 percent of total world electricity generation in 2014, it's growing rapidly, and it provides significantly higher amounts in some jurisdictions, demonstrating its potential. Denmark got 41.4 percent of its power from wind in 2014, and Spain, Portugal, Ireland, and Lithuania all got more than 15 percent, with Spain reaching up to 59.6 percent during peak periods. China's wind power market grew by 45 percent in 2014.[27] The Global Wind Energy

Council reports that, with close to $100 billion in investments globally, wind is "one of the fastest growing industrial segments in the world."[28] Still, people complain that wind power installations are harmful to human health, kill birds and other wildlife, and are unsightly.

When I think about developments and controversy around wind power, I often think of my cabin on Quadra Island off the British Columbia coast. It's as close to my heart as you can imagine. From my porch I can see clear across the waters of Georgia Strait to the snowy peaks of the rugged Coast Mountains. It's one of the most beautiful views I have seen. And I would gladly share it with a wind farm.

Sometimes it seems I'm in the minority. Across Europe and North America, environmentalists and others are locking horns with the wind industry over farm locations. In Canada, opposition to wind installations has sprung up from Nova Scotia to Ontario to Alberta to B.C. In the U.K., more than one hundred national and local groups, led by some of the country's most prominent environmentalists, have argued wind power is inefficient, destroys the ambience of the countryside, and makes little difference to carbon emissions. And in the U.S., the Cape Wind Project, which would site 130 turbines off the coast of affluent Cape Cod, Massachusetts, has come under fire from famous liberals, including John Kerry and the late senator Edward Kennedy.

It's time for some perspective. With the growing urgency of climate change, we can't have it both ways. We can't shout about the dangers of global warming and then turn around and shout even louder about the "dangers" of windmills. Climate change is one of the greatest challenges humanity will face this century. Confronting it will take a radical change in

the way we produce and consume energy—another industrial revolution, this time for clean energy, conservation, and
efficiency.

We've undergone such transformations before and we can
again. But we must accept that all forms of energy have associated costs. Fossil fuels are limited in quantity, create vast
amounts of pollution, and contribute to climate change. Large-
scale hydroelectric power floods valleys and destroys valuable
farmland and wildlife habitat. Nuclear power plants are expensive, create radioactive waste, and take a long time to build.

Wind power's critics aren't always entirely wrong. It does
have its downsides. It's highly visible and can kill birds. But any
human-made structure (not to mention cars and house cats)
can kill birds—including houses, radio towers, and skyscrapers.
In Toronto alone, an estimated 1 million birds collide with the
city's buildings every year.[29] In comparison, the risk to birds
from well-sited wind farms is low. Even the U.K.'s Royal Society for the Protection of Birds says scientific evidence shows
wind farms "have negligible impacts" on birds when they are
appropriately located.[30] New technologies also reduce the
numbers of birds and bats killed by wind installations. Fossil
fuels, especially coal, are far more dangerous to birds, bats, and
other wildlife. Heavy metals such as mercury and lead from
burning coal kill numerous birds and can affect their ability to mate and protect territory by changing the ways they
communicate. Climate change is also affecting many species'
breeding and migratory patterns. Analysis by U.S. News & World
Report in 2014 on the numbers of birds killed every year by U.S.
electricity sources found that between 1,000 and 28,000 are
killed by solar installations; between 140,000 and 328,000 by
wind power; about 330,000 by nuclear; 500,000 to 1 million

by oil and gas; and 7.9 million by coal.[31] Between 12 million and 64 million birds a year are also killed in the U.S. by transmission lines. Domestic cats are the biggest danger, though, killing from 1.4 to 3.7 billion a year!

Improved technologies and more attention to wind farm placement can clearly reduce harm to birds, bats, and other wildlife. Indeed, the real risk to flying creatures comes not from windmills but from a changing climate, which threatens the very existence of species and their habitats. Wind farms should always be subject to environmental-impact assessments, but a blanket "not in my backyard" approach is hypocritical and counterproductive.

Other arguments against wind power focus on potential detrimental effects on human health. Leaving aside the well-known fact that burning fossil fuels causes far more human health problems and deaths than wind or any other form of clean energy, are these arguments valid? There's no doubt that wind turbines, especially older ones, can be noisy, and some people find them unsightly. But many problems can be addressed by locating quieter turbines far enough from human habitation to reduce impacts.

As for health effects, a comprehensive 2014 Health Canada study confirmed previous research: although people report being annoyed by wind turbines, there's no measurable association between wind turbine noise and sleep disturbance and disorders, illnesses, and chronic health conditions, or stress and quality-of-life issues.[32] An independent panel convened by the Massachusetts Department of Environmental Protection in 2012 reviewed the available research and found no scientific evidence to support most claims about "wind turbine syndrome": infrasound effects and harm blamed on wind power such as pain and stiffness, diabetes, high blood pressure, tinnitus, hearing impairment, cardiovascular disease, and headache/migraine.[33]

At worst, there is some evidence that wind installations may cause some annoyance and sleep disruption. But most of the resulting minor effects can be overcome by regulations governing how close windmills are to residences. In Ontario, the required setback is 550 metres. At this distance, audible sound from windmills is normally below forty decibels, which is about what you'd find in most bedrooms and living rooms. A 2013 Australian report concluded people living near wind installations where anti-wind campaigns were active were more likely to report health problems, suggesting some issues may be psychological.[34] And, for a 2014 study published in the American Psychological Association's *Health Psychology* journal, researchers from the University of Auckland showed readily available anti-wind film footage to twenty-seven people. Another twenty-seven were shown interviews with experts who said infrasound, such as that created by wind turbines, can't directly cause negative health effects. Subjects were then told they would be exposed to two ten-minute periods of infrasound, but were actually only exposed to one. After both real and "sham" exposure, people in the first group were far more likely to report negative symptoms than those in the second. In fact, subjects in the second group reported "no symptomatic changes" after either exposure. According to the researchers, "Results suggest psychological expectations could explain the link between wind turbine exposure and health complaints."[35]

After its study, Health Canada cautioned that more research may be needed and we shouldn't downplay the annoyance factor. Again, improvements in technology and proper siting will help overcome many problems. And there's no doubt that fossil fuel development and use—from bitumen mining, deep-sea drilling, mountaintop removal, and fracking to wasteful burning in single-user vehicles and power plants—are far more annoying and damaging to human health than wind power and other renewable energy technologies.

As for the question of whether windmills blight landscapes with their unsightliness, Mostafa Tolba, executive director of the UN Environment Programme from 1976 to 1992, told me belching smokestacks were considered signs of progress when he was growing up in Egypt. Even as an adult concerned about pollution, it took him a long time to get over the pride he felt when he saw a tower pouring clouds of smoke. Our perception of beauty is shaped by our values and beliefs. Some people think wind turbines are ugly. I think smokestacks, smog, acid rain, coal-fired power plants, and climate change are ugly. I think windmills are beautiful. They harness the wind's power to supply us with heat and light. They provide local jobs. They help clean air and reduce climate change.

Pursuing wind power as part of our move toward clean energy makes sense. Wind power has become the fastest-growing source of energy in the world, employing hundreds of thousands of workers. That's in part because larger turbines and greater knowledge of how to build, install, and operate them has dramatically reduced costs over the past two decades. Prices are now comparable to other forms of power generation and will likely decrease further as technology improves.

If one day I look out from my cabin porch and see a row of windmills spinning in the distance, I won't curse them. I will praise them. It will mean we're finally getting somewhere.

## The Future of Hydro in a Warming World

PEOPLE HAVE HARNESSED energy from moving water for thousands of years. More than two thousand years ago, Greek farmers used various types of water wheels to grind grain in mills. In the late 1800s, people figured out how to harness the power to produce electricity. Throughout the twentieth century and into the twenty-first, hydropower has expanded, producing about 17 percent of the world's electricity by 2014 and about 85 percent of renewable energy—and it shows no signs of slowing.[36]

According to the online magazine *WaterWorld* in 2014, "An expected 3,700 major dams may more than double the total electricity capacity of hydropower to 1,700 GW within the next two decades"[37]—including in my home province of B.C., where the government has started a third dam on the Peace River at Site C. "Hydropower is the most important and widely-used renewable source of energy," the U.S. Geological Survey says.[38] How "green" is hydropower, and how viable is it in a warming world with increasing water fluctuations and shortages? To some extent, it depends on the type of facility.

The Center for Climate and Energy Solutions notes some large dams are used mainly for purposes such as water storage or flood control with power generation an additional function, whereas some are used primarily to generate electricity.[39] Small hydro such as run of river is installed on running water and doesn't use water stored in reservoirs. Pumped storage facilities don't generate additional energy. Rather, they store energy by pumping water from a lower reservoir to a higher one when demand and price are low, potentially using renewable energy sources, and releasing water through turbines when price and demand are high. All have varying environmental impacts.

One of the biggest trade-offs with large-scale hydro facilities is that building them means flooding land that is often valuable for agriculture and human habitation. Damming rivers also impedes fish—even with technologies like fish ladders—and can have negative impacts on wildlife habitat, as well as altering river temperatures, dissolved oxygen levels, and flows. And while hydropower creates fewer pollution and climate impacts than fossil fuel power plants, it isn't entirely clean. Greenhouse gases can be released as vegetation is cleared to build a dam and flood land, and methane can build up and be released from reservoirs as vegetation decays and water levels fluctuate.

Ironically, although hydropower is seen as an energy source that helps slow global warming, its viability in many areas is threatened by climate change. As greenhouse gas emissions rise and the world warms, the entire hydrologic cycle—surface and ground water, glaciers, precipitation, runoff, and evaporation—are all affected. Altered precipitation and increased droughts are altering water levels in rivers and behind hydro dams. The massive Hoover Dam on the Colorado River is now operating at 30 percent capacity, and new turbines have to be installed at lower elevation, because of low precipitation and drought. In Nepal, "low water levels rendered a brand-new dam project ineffective and cut off the water supply farther downstream," according to John Matthews, director of freshwater and adaptation at Conservation International.[40]

Matthews and coauthors of a study in the journal PLoS Biology wrote that, according to the Organisation for Economic Co-operation and Development, climate change puts

40 percent of hydro development investments at risk.[41] They recommend a different approach to dams and hydro that takes climate change into account, by building projects in stages so that adjustments can be made as more is known about climate patterns, or by "building with nature" rather than on top of it.

Meanwhile, as more environmentally benign power technologies become increasingly cost effective and viable, the U.S. is removing older dams, many of which don't even have fish ladders, because costs to maintain and repair them are too high.

Various types of hydropower will remain a part of the clean energy equation, but there's a need to find the least disruptive, most efficient methods. As scientist Peter Gleick, president and cofounder of California's Pacific Institute, notes, the key to supplying energy to growing populations in a warming world will be to use a diversity of power sources. "We need to design our energy systems to be resilient in the face of growing uncertainty about technology and climate and national security and all of the factors that affect energy."[42]

## Storing Energy Makes Renewables Doable

REMOTE COMMUNITIES IN Australia often use diesel generators for power. They're expensive to run and create lots of pollution and greenhouse gas emissions. Even people who don't rely entirely on diesel generators use Australia's power grid, mostly fuelled by burning coal, which also pollutes and creates climate-altering emissions. An indigenous company is working to change that, and to show that getting energy needn't be expensive or polluting.

AllGrid Energy is producing ten-kilowatt-hour solar power batteries to take advantage of Australia's abundant sunlight and consequent demand for solar panels.[43] Tesla is also selling its Powerwall home battery systems in Australia. The AllGrid uses materials such as lead acid gel that make it less expensive than the Powerwall, which uses lithium. Many of AllGrid's systems are being sold in remote indigenous communities, giving them new levels of energy independence at affordable prices.

It's just one example of the continuing rapid advances in renewable energy technology—one that clears a previous hurdle with many renewable energy technologies: their intermittent nature. Fossil fuels are both energy sources and storage systems. Renewable energy, such as wind and solar, is only a source.

Many people have argued that because solar and wind energy only work when the sun shines or the winds blow, and output varies according to wind speed, cloud cover, and other factors, they can never replace large "baseload" sources such as coal, oil, gas, and nuclear. But batteries and other energy storage methods, along with new developments in power grids, make renewables competitive with fossil fuels and nuclear power—and often better—in terms of reliability, efficiency, and affordability.[44]

The Tesla Powerwall and AllGrid systems also address another barrier to renewable energy: cost. Both are designed to be relatively affordable, and prices for storage are dropping as the technology improves. The World Energy Council forecast in 2016 that costs for energy storage could drop by 70 percent in fifteen years, with costs for energy production from renewable sources such as wind and solar also dropping. According to the WEC, "Affordable, efficient storage would enable volatile renewable energies to be better integrated into electricity systems; it would greatly improve the economics of home solar systems, even allowing people to go off grid; and it could help catalyze a revolution in electric cars as well as help solve the problems of the grid going down or grid overload."[45] The council's 2016 report, E-Storage: Shifting from Cost to Value, also found that current cost estimates for energy storage are exaggerated because they fail to fully take into account the value they bring to some situations, such as the ability to store power during windy or sunny weather and sell it back when demand and costs rise.[46]

A Stanford University study found that wind power in the U.S. is already developed to a point that it could be sustainable with good storage systems but that because of the extra energy required to produce photovoltaic cells, solar still has a way to go.[47] With rapidly advancing storage technologies, renewable energy could easily replace dangerous fossil fuels in fairly short order—and the Stanford study concluded that the U.S. could get its power entirely from renewables without storage by making the grid more efficient.[48]

Ontario's Independent Electricity System Operator, or IESO, has contracted five companies to test and demonstrate a number of storage systems, including various types of batteries, hydrogen storage, kinetic flywheels, and thermal systems that store heat in special bricks. "Energy storage projects will provide more flexibility and offer more options to manage the system efficiently," IESO president Bruce Campbell told the Globe and Mail in 2014.[49] According to the Union of Concerned Scientists, the main storage methods include thermal, compressed air, hydrogen, pumped hydroelectric, flywheels, and batteries.[50]

Thermal energy storage systems heat or cool a storage medium, such as water, salt, or chemicals. Techniques include sensible, latent, and thermochemical storage. Sensible is the simplest and oldest, and includes hot water tanks, from individual home water heaters to district heating systems. "Sensible" simply refers to changing the temperature of the storage medium, which can be water, rock, salt, sand, or earth. One way that solar energy can be stored is to heat water with solar thermal panels on a roof, and pump and store the heated water in the ground or insulated water tanks. Latent storage is more efficient, as less of the energy is lost in the process and storage. This refers to a process whereby the storage medium changes form, such as by melting or solidifying, absorbing, or releasing energy as the material changes. Molten salt is often used. Thermochemical means storing heat in chemical bonds. Salt hydration is a common method.

Compressed air storage takes air and compresses and stores it, often in an underground cavern or depleted natural gas well, and then heats and expands it in an expansion turbine to drive a power generator. Several methods are available, including diabatic and adiabatic, meaning involving the transfer of heat, or not.

With hydrogen storage, electrolysis—using direct electric currents to create chemical reactions—is employed to convert electricity to hydrogen, which is then stored in pressurized vessels, solid metal hydrides or nanotubes, or underground salt caverns, and later converted back to electricity. It can be electrified with fuel cells or burned in power plants.

With pumped hydroelectric, water is pumped during periods when energy demand is low from a lower reservoir to a higher one, then released during high demand through turbines, as in a conventional hydro dam.

Flywheel systems use kinetic energy by putting a spinning rotor in an enclosure with little friction. Electricity is used to make the flywheel spin at maximum speed. Because of the lack of friction, inertia allows the rotor to keep spinning. According to the Energy Storage Association, "energy is discharged by drawing down the kinetic energy," using an integrated motor-generator that was also used to make the rotor spin.[51]

Batteries may be the most well-known way to store energy. Batteries don't store electrical energy; rather, they store energy in chemical form, and convert the chemical energy to electrical energy when it is required. Although there are many different types of battery, most contain an electrochemical cell or cells, consisting of two separated electrodes.

Battery storage employs a number of techniques. Lead acid, nickel cadmium, lithium ion, and sodium sulphur batteries are all in use, and have varying advantages and disadvantages. Some have short life cycles and some can overheat easily. Costs for materials used to produce them can vary as well. Industrial-scale flow batteries store charges in liquid electrolytes that are "then pumped through an electrode assembly, known as a stack, containing two electrodes separated by an ion-conducting membrane," according to Science online.[52] The tanks used to hold the liquids can be any size, so they can be used to store large amounts of energy. Different types of flow batteries have been developed, including redox, hybrid, and membraneless. According to the Science article, by combining liquid flow technology with solid lithium storage materials, researchers were able to store ten times more energy per volume in the tanks. One advantage of flow batteries is that they can be recharged instantly by replacing liquid in the tanks.

Harvard researchers have been developing a power grid–scale flow battery that uses organic compounds called quinones instead of expensive precious metals. Quinones are found in plants and can be synthesized from oil. The main advantage is that using abundant quinones brings the costs down substantially. "A safe and economical flow battery could play a huge role in our transition off fossil fuels to renewable electricity," lead researcher Michael J. Aziz said.[53]

On an individual scale, batteries like the AllGrid and Tesla's Powerwall allow homeowners to store energy from solar panels or from the grid when utility rates are low to power the home in the evening. According to Tesla, "Each Powerwall has a 6.4 kWh energy storage capacity, sufficient to power most homes during the evening using electricity generated by solar panels during the day. Multiple batteries may be installed together for homes with greater energy needs."[54]

As electric cars become more popular, their batteries could even be connected to grids to supply excess power, which could even offset costs for car owners.

Renewable energy with storage has a number of advantages over fossil fuels. It can discharge power to the grid much more quickly to meet demand, and it can offer more reliable power to communities that aren't near the transmissions system. It is also more reliable and less prone to disruption, because the power sources are distributed over a large area, so if one part is knocked out by a storm, for example, other parts will keep the system running. Many fossil fuel and nuclear power systems also require a lot of water for cooling and so can be affected by drought, whereas wind and solar power don't require water.

Given the rapid pace of all kinds of technological development, storage methods are likely to help renewable sources clear a number of hurdles over the coming years. But renewables are already being deployed without storage, and a 2016 study showed the U.S. could reduce $CO_2$ emissions from its electricity sector by 80 percent relative to 1990 levels within fifteen years "with current technologies and without electrical storage."[55]

The study, by scientists from the National Oceanic and Atmospheric Administration and University of Colorado Boulder, and published in *Nature Climate Change*, concluded that grid improvements, including a new high-voltage direct-current transmission grid, could deliver low-cost clean energy throughout the country to match supply and demand.[56]

Still, storage has many advantages, and a perceived lack of demand could slow development of new technologies. With the urgent need to cut greenhouse gas emissions, governments need to provide incentives for rapid renewable energy development and deployment. The main barriers are more political than technological.

## Geothermal: Tapping Earth's Abundant Energy

IN 2014, AMID the controversy over B.C.'s Peace River Site C dam project, the Canadian Geothermal Energy Association released a study showing the province could get the same amount of energy more affordably from geothermal sources for about half the construction costs.[57] Unlike Site C, geothermal wouldn't require massive transmission upgrades, would be less environmentally disruptive, and would create more jobs throughout the province rather than in just one area.

Despite the many benefits of geothermal, Canada is the only Pacific Ring of Fire country that doesn't use it for commercial-scale energy. According to DeSmogBlog, "New Zealand, Indonesia, the Philippines, the United States and Mexico all have commercial geothermal plants."[58] Iceland heats up to

90 percent of its buildings and homes, as well as swimming pools, with geothermal.

Geothermal energy is generated by heat from Earth's rocks, liquids, and steam. It can come from shallow ground, where the temperature is a steady ten to sixteen degrees Celsius; hot water and rocks deeper in the ground; or possibly very hot molten rock (magma) deep below Earth's surface. As with clean energy sources like solar, geothermal energy systems vary, from heat transfer systems that use hot water from the ground directly to heat buildings, greenhouses, and water, to power-generating systems that pump underground hot water or steam to drive turbines. The David Suzuki Foundation's Vancouver and Montreal offices use ground-source heat pumps for heating and cooling, which are different from geothermal systems that generate power, as they use liquid pumped through underground to heat and cool the buildings.

According to *National Geographic*, geothermal power plants use three methods to drive turbines. Dry steam uses steam from fractures in the ground. "Flash plants pull deep, high-pressure hot water into cooler, low-pressure water," which creates steam. In binary plants, which produce no greenhouse gas emissions and will likely become dominant, "hot water is passed by a secondary fluid with a much lower boiling point," which turns the secondary fluid into vapour.[59]

Unlike wind and solar, geothermal provides steady energy and can serve as a more cost effective and less environmentally damaging form of baseload power than fossil fuels or nuclear. It's not entirely without environmental impacts, but most are minor and can be overcome with good planning and siting. Geothermal fluids can contain gases and heavy metals, but most new systems recycle them back into the ground. Operations should also be located to avoid mixing geothermal liquids with groundwater and to eliminate impacts on nearby natural features such as hot springs. Some geothermal plants can produce small amounts of $CO_2$, but binary systems are emissions free. In some cases, resources that provide heat can become depleted over time.

Geothermal potential has been constrained by the need to locate operations in areas with high volcanic activity, geysers, or hot springs. A 2015 Stanford University study of U.S. renewable energy potential concluded, "Geothermal energy was available at a reasonable cost for only 13 states."[60] But new developments are making it more widely viable. One controversial method being tested is similar to fracking for oil and gas.[61] Water is injected into a well with enough pressure to break rock and release heat to produce hot water and steam to generate power through a turbine or binary system.

Researchers have also been studying urban heat islands as sources of geothermal energy.[62] Urban areas are warmer than their rural surroundings, both above and below ground, because of the effects of buildings, basements, and sewage and water systems. Geothermal pumps could make the underground energy available to heat buildings in winter and cool them in summer.

New methods of getting energy from the ground could also give geothermal a boost. Philanthropist and entrepreneur Manoj Bhargava is working with researchers to bring heat to the surface using graphene cords rather than steam or hot water.[63] Graphene is stronger than steel and conducts heat well. Bhargava says the technology would be simple to develop and employ and could be integrated with existing power grids.

Unfortunately, geothermal hasn't received the same level of government support as other sources of energy, including fossil fuels and nuclear. That's partly because upfront costs are high, and as with oil and gas exploration, geothermal sources aren't always located where developers hope they'll be. As DeSmog notes, resources are often found in areas that already have access to inexpensive hydropower.[64]

Rapid advancements in renewable energy and power-grid technologies could put the world on track to a mix of clean sources fairly quickly—which is absolutely necessary to curtail global warming. Geothermal energy can and should be part of that mix.

## Banking on Biofuels

MUCH OF THE shift from fossil fuels to renewable energy is taking place at the power plant level, where electricity is transmitted through grids for use in homes and buildings. What about transportation fuels? Can we make necessary reductions in greenhouse gas emissions by switching to cleaner fuels? Or is this just an attempt to keep our twentieth-century technology chugging along, while trading one set of environmental problems for another?

Many energy studies conclude that biofuels will play a role in an extremely low-carbon energy system. For some uses, like freight and aviation, few alternatives exist. These vehicles need energy-dense liquid fuels to operate. If we are to reduce our greenhouse gas emissions by the 80 percent required to avoid two degrees Celsius of warming, we will have to consider the most environmentally and socially friendly biofuels among the available energy options. The challenge is to minimize the ecological and food-system impacts.

Biofuels aren't new, nor are they used solely as transportation fuels.[65] Wood can be burned in power plants, for example, and many of the first cars, including Ford's Model T, were run on ethanol or peanut oil. But they're now seen as a way to cut down on burning fossil fuels for transportation. In fact, all biofuels use solar energy that has been converted to chemical energy through photosynthesis, so in a sense, even fossil fuels are a form of biofuel. But because they're based on "ancient" photosynthesis, they are considered nonrenewable, unlike biofuels that use current photosynthesis.

Renewable biofuels offer a number of advantages over fossil fuels. Most are less toxic. The crops used to produce them can be grown quickly, so unlike coal, oil, and gas that take millions of years to form, they're considered renewable. They can also be grown almost anywhere, reducing the need for infrastructure such as pipelines and oil tankers, and conflicts around scarcity and political upheaval in many areas where fossil fuels are found.

The main biofuels are ethanol and biodiesel, but biomass like wood can be burned directly for fuel, though that usually creates more greenhouse gas emissions than burning fossil fuels to produce the same amount of energy.[66] Greenhouse gas emissions from biofuels are offset to a great extent because the plants absorb and store carbon dioxide while they're growing, so the $CO_2$ emissions are roughly equal to the amount of $CO_2$ stored by the crops.

Despite the advantages, numerous problems with biofuels have led many to question whether or not they can truly be considered a green alternative.[67] Andrew Steer and Craig Hanson of the World Resources Institute noted in a 2015 *Guardian* article that biofuel has three major strikes against it: "It uses land needed for food production and carbon storage, it requires large areas to generate just a small amount of fuel, and it won't typically cut greenhouse gas emissions."[68]

According to a 2012 study by the European Academies Science Advisory Council, when all the impacts of "first-generation" crop-based biomass cultivation and production are taken into account, overall greenhouse gas emissions reductions are minimal or nonexistent, and food, agriculture, and natural ecosystems are put at risk.[69]

Using crops like corn to produce biofuel often requires converting land from food production to fuel production or destroying natural ecosystems that provide valuable services, including carbon sequestration. Crops also require fertilizers, pesticides, and large amounts of water, as well as machinery for planting, growing, harvesting, transporting, and processing. If forests are cleared to make way for fuel crops, and if the entire life cycle of the fuels is taken into account, biofuels don't always reduce the overall amount of greenhouse gases pumped into the atmosphere. Palm oil, used for biodiesel, is especially bad, because valuable carbon sinks such as peat bogs or rain forests are often destroyed to grow palms.[70] The European Union has moved to limit production and use of biofuels that compete with food crops, and may ban these first-generation biofuels altogether.

Using better farming methods and more efficient feed-stocks, and growing fuel crops on land that isn't good for growing food, can reduce impacts on land use and climate.[71] For example, fast-growing grasses, agricultural and forestry wastes, and even household wastes can be used rather than crops such as corn that are normally considered food. Some feedstocks are more efficient at producing energy than others, even among first-generation stocks. Canola and sugarcane deliver a greater percentage of energy compared to the energy required to produce fuel than corn.

Biofuels that are grown for fuel production on lands that aren't suitable for food production, inedible parts of food crops, and forest wastes are known as second-generation biofuels. These normally use cellulosic materials, including switch grass and agricultural and forestry wastes, which are more efficient and produce fewer greenhouse gases, and don't normally displace food crops. However, the process of converting cellulose to ethanol is more difficult than turning sugars from corn or sugarcane to fuel. Cellulose is the structure that keeps plant cells together. To produce fuel, the cellulose must be broken down using enzymes to get sugar, which is then converted to ethanol or biodiesel. This process is expensive, but research is being conducted into more efficient ways to break down cellulose and refine fuels. Some studies show these biofuels can produce 540 percent more energy than that required to produce the fuel, compared to just 25 percent more for corn-based ethanol.

So-called third-generation biofuels made from aquatic microalgae show a great deal of promise, as they're efficient and can be produced without large land bases. They grow quickly and can be harvested every one to ten days. Because algae also require carbon dioxide to grow, algae used to produce fuel can even be used to capture and store $CO_2$ from power plants. But because algae have low photon-to-fuel conversion efficiency, and require a lot of nutrients, water, and energy to grow and convert to fuel, they are not yet economically viable. But their high oil content and the fact that they can be grown without using valuable agricultural land have created a lot of interest, leading to considerable investment and research, especially in the U.S. Harmful algal blooms and waste-water can even provide the basis for biofuels made from algae.

Fourth-generation biofuels, which use synthetic biology tools, are still in the experimental stage. Many of these technologies focus on designing organisms such as bacteria through genetic manipulation to take in $CO_2$ and release sugars that can be converted to fuel. Electrofuels, which convert $CO_2$ into organic compounds without using photosynthesis, are another fourth-generation biofuel under study. Researchers at the U.S. Lawrence Berkeley National Laboratory are experimenting with microbes that convert $CO_2$ into fuel using hydrogen and electricity.[72]

Biodiesel and gasoline mixed with ethanol are already widely available. Biofuels can play an important role in reducing greenhouse gas emissions, especially for applications such as long-haul trucking, but the massive amounts of land, biomass, and water required to produce current biofuels mean they aren't a panacea. We can make greater gains in the transportation sector by focusing on fuel efficiency and conservation, increased public transit, and other alternatives to private automobiles, and shifting to electric vehicles, especially as clean electricity sources become more widely available. But research into new types of biofuels may provide ways to move trucks, trains, and airplanes with fewer environmental and climate consequences.

### Trash Incineration: Viable Energy Source or Waste of Resources?

MANY URBAN AREAS have built or are considering building waste-incineration facilities to generate energy. At first glance, it seems like a win-win. You get rid of "garbage" and acquire a new energy source with fuel that's almost free. But it's a problematic solution, and a complicated issue.

Metro Vancouver has a facility in Burnaby and is planning to build another, and Toronto is also looking at the technology, which has been used elsewhere in the region, with a plant in Brampton and another under construction in Clarington.[73] The practice is especially popular in the European Union, where countries including Sweden and Germany now have to import waste to fuel their generators.

The term "waste" is correct; there's really no such thing as garbage. And that's one problem with burning it for fuel. Even those who promote the technology would probably agree that the best ways to deal with waste are to reduce, reuse, and recycle it. It's astounding how much unnecessary trash we create, through excessive packaging, planned obsolescence, hyper-consumerism, and lack of awareness. This is one area where individuals can make a difference, by refusing to buy over-packaged goods and encouraging companies to reduce packaging, and by curbing our desire to always have newer and shinier stuff.

We toss out lots of items that can be reused, repaired, or altered for other purposes. As for recycling, we've made great strides, but we still send close to three-quarters of our household waste to the landfill. People in countries like Canada are especially wasteful—each Canadian produces around 900 kilograms of waste a year.[74] Much of the material that ends up in landfills is usable, compostable, or recyclable, including tons of plastics.

Turning unsorted and usable trash into a valuable fuel commodity means communities are less likely to choose to reduce, reuse, and recycle it.[75] Burning waste can seem easier and less expensive than sorting, diverting, and recycling it. But once it's burned, it can never be used for anything else—it's gone!

Incinerating waste also comes with environmental problems.[76] Although modern technologies reduce many air pollutants once associated with the process, burning plastics and other materials still creates emissions that can contain toxins such as mercury, dioxins, and furans. As with burning fossil fuels, burning waste—much of which is plastics derived from fossil fuels—also produces carbon dioxide and nitrous oxide emissions that contribute to climate change.[77]

Burning waste doesn't make it disappear, either. Beyond the fly ash and pollutants released into the atmosphere, a great deal of toxic bottom ash, or noncombustible residue, is left over. Metro Vancouver says bottom ash from its Burnaby incinerator is about 17 percent the weight of the waste burned. That ash must be disposed of, usually in landfills. Metro testing has found high levels of the carcinogenic heavy metal cadmium in bottom ash, sometimes twice the limit allowed for landfills. High lead levels have also been reported.

Incineration is also expensive and inefficient. Once we start the practice, we come to rely on waste as a fuel commodity, and it's tough to go back to more environmentally sound methods of dealing with it. As has been seen in Sweden and Germany, improving efforts to reduce, reuse, and recycle can actually result in shortages of waste "fuel"!

It's a complicated issue. We need to find ways to manage waste and to generate energy without relying on diminishing and increasingly expensive supplies of polluting fossil fuels. Sending trash to landfills is clearly not the best solution. But we have better options than landfills and incineration, starting with reducing the amount of waste we produce. Through education and regulation, we can reduce obvious sources and divert more compostable, recyclable, and reusable materials away from the dump. It's simply wasteful to incinerate it.

It would be far better to sort trash into organics, recyclables, and products that require careful disposal. We could then divert these different streams to minimize our waste impacts and produce new commodities. Organics could still be incinerated in biomass energy systems to help offset fossil fuel use, while creating valuable supplies of fertilizers. Diversion and recycling lessen the need to extract new resources and disrupt the environment, while creating more value and jobs. That's a win all around!

## Carbon Sinks and Natural Capital

ALTHOUGH MUCH OF the debate and action around climate change has focused on curbing emissions from burning fossil fuels such as oil, coal, and gas, the destruction of forests, wetlands, grasslands, and peatlands is responsible for about one-quarter of all other emissions into the atmosphere. That's higher than emissions from cars, trucks, boats, and planes together.

Throughout the world, forests are being rapidly cleared for agriculture and oil and gas development and are being destructively mined and logged. The destruction of forests and other ecosystems is not only a driver of extinction of species, such as boreal caribou, but of global warming as well.

A 2009 report published in the journal Nature shed some light on the role of forests, grasslands, soils, and other ecosystems in the climate equation.[78] We've known for a long time that forests are important carbon sinks that absorb carbon from the atmosphere and prevent it from contributing to global warming. The *Nature* study showed that tropical forests absorb more carbon than we realized. Researchers from a number of institutions, including the University of Toronto, analyzed data from 79 intact forests in Africa from 1968 to 2007, along with similar data from 156 intact forests from twenty non-African countries. They concluded that tropical forests absorb about 4.8 billion metric tons of carbon a year, equivalent to about 18 percent of the carbon dioxide added to the atmosphere each year by burning fossil fuels. The world's oceans are the other major carbon sink, absorbing about half the human-produced carbon that doesn't end up in the atmosphere.

That doesn't mean we can count on the forests or the oceans to save us from our folly. To start, about 15 billion of the 32 billion metric tons of carbon dioxide that humans produce is not reabsorbed on land or sea and ends up in the atmosphere. And the carbon stored in forests can be released back into the atmosphere with natural disturbances, such as fire or insect outbreaks, or if the forest is logged. This is because when trees are cut down, die, and decay naturally, or burn, some of the stored carbon is released back into the atmosphere. Even if carbon remains stored in the form of wood products taken from a logged forest, some is still released when logging disturbs soils in the forest floor. And many wood and pulp and paper products are discarded and destroyed in a much shorter time period than the life of an old-growth forest. This means that the carbon is released earlier than it would have been if the forests were left intact.

We humans have upset the balance of nature in more ways than we understand. Scientists who worked on the *Nature* study haven't figured out why the tropical trees are growing big enough to absorb more carbon than they release. One theory is that global warming and the extra carbon in the atmosphere are actually fertilizing the trees. One thing we do know is that we cannot rely on tropical forests to prevent dangerous levels of climate change. But the amount of carbon they store gives us another compelling argument for protecting forests, as they may at least provide a buffer while we work on other solutions

Clearly, it's not the only reason to protect forests. Looking at the ability of forests to absorb carbon allows us to see that they have economic value beyond resources such as lumber that we have traditionally considered. Forests are a source of medicine, food, and clean drinking water, and are habitat for more than half of all land-based plants and animals on the planet. Forests also provide spiritual, aesthetic, and recreational opportunities for millions of people. Forest degradation is also contributing to another ecological crisis, a biodiversity crisis on par with earlier mass extinctions. Scientists estimate that sixteen thousand species are now threatened with extinction, including 12 percent of birds, 23 percent of mammals, and 32 percent of amphibians. Habitat destruction is partly responsible for this crisis, and climate change is exacerbating it. And although most of our carbon emissions are from burning fossil fuels, one-quarter is from deforestation.

This shows that everything in nature is interconnected, and the planet works to find equilibrium. We can't confront the problems we have created on a piecemeal basis. We must look at them together. Conserving the world's forests—which can include sustainable forestry practices—is one obvious place to start dealing with some of the most imminent crises.

We need to adopt a carbon stewardship approach to how we use our forests and the goods and services we take from them. For some scientists, this means setting aside at least half of all remaining intact forests as protected areas, particularly carbon-rich forests like old-growth temperate rain forests in B.C. and the boreal in Canada's north, where wildlife like caribou feed, breed, and roam. Protecting intact forests also promotes ecological resiliency so that species and ecosystems can cope with and adapt to the effects of climate change. That doesn't mean logging companies should be allowed to trash the other 50 percent. Forests that we do manage for wood and paper production should be logged according to the highest standards of ecosystem-based management, without clear-cutting, and with adequate protection for wildlife habitat, as well as sensitive areas like wetlands.

Besides helping to prevent global warming, green spaces and natural features can also protect against its consequences, including floods, drought, and fires. Recent floods in Canada, Europe, India, the U.S., and elsewhere were previously referred to as "once in a generation" or "once in a century." Now, they're scaled up to "once in a decade." Scientists and insurance executives alike predict extreme weather events will increase in intensity and frequency. Climate change is already having a dramatic impact on the planet. Communities around the world are rallying to prepare.

Although calls are mounting for the need to rebuild and strengthen infrastructure such as dikes, storm-water management systems, and stream-channel diversion projects, we've overlooked one of our best tools for fighting climate change: nature. By protecting nature, we protect ourselves, our communities, and our families. The business case for maintaining and restoring nature's ecosystems is stronger than ever. Wetlands, forests, flood plains, and other natural systems absorb and store water and reduce the risk of floods and storms, usually more efficiently and cost effectively than built infrastructure. Wetlands help control floods by storing large amounts of water during heavy rains—something paved city surfaces just don't do.

A study of the Upper Mississippi and Missouri Basins showed wetland restoration would have provided enough floodwater storage to accommodate excess river flows associated with flooding in the U.S. Midwest in 1993.[79] Research done for the City of Calgary more than thirty years ago made similar suggestions about the value of protecting flood plains from overdevelopment.[80] When wetlands are destroyed, the probability of a heavy rainfall causing flooding increases significantly. Yet we're losing wetlands around the world at a rate estimated at between 1 and 3 percent a year.[81]

By failing to work with nature in building our cities, we've disrupted hydrologic cycles and the valuable services they provide. The readily available benefits of intact ecosystems must be replaced by human-made infrastructure that can fail and is costly to build, maintain, and replace.

Protecting and restoring rich forests, flood plains, and wetlands near our urban areas is critical to reduce carbon emissions and protect against the effects of climate change. Nature effectively sequesters and stores carbon, helping to reduce greenhouse gas emissions. It also regulates water. Forested basins, for example, have greater capacity to absorb water than clear-cut areas where higher peak stream flows, flooding, erosion, and landslides are common.

How can we protect ecosystems rather than seeing conservation as an impediment to economic growth? The answer is to recognize their real value. The David Suzuki Foundation has evaluated some of Canada's natural assets.[82] This approach calculates the economic contribution of natural services, such as flood protection and climate regulation, and adds that to our balance sheets. Because traditional economic calculations ignore these benefits and services, decisions often lead to the destruction of the very ecosystems upon which we rely. Unfortunately, we often appreciate the value of an ecosystem only when it's not there to do its job.

Cities around North America are discovering that maintaining ecosystems can save money, protect the environment, and create healthier communities. A study of the Bowker Creek watershed on southern Vancouver Island showed that by incorporating rain gardens, green roofs, and other green infrastructure, peak flows projected for 2080 from increased precipitation because of climate change could be reduced by 95 percent.[83] Opting to protect and restore watersheds in the 1990s rather than building costly filtration systems has saved New York City billions of dollars.[84]

Intact ecosystems are vital in facing the climate change challenges ahead. They also give us health and quality-of-life benefits. Responsible decision-making needs to consider incentives for protecting and restoring nature, and disincentives for degrading it.

## Do We Really Have to Go Nuclear?

AS KNOWLEDGE ABOUT climate change increases, so does demand for clean energy. But many argue that, despite the 2011 Fukushima nuclear disaster and others like Chernobyl and Three Mile Island, nuclear is the best option to reduce carbon emissions fast enough to avoid catastrophic climate change.

Conventional nuclear technology still comes with a lot of problems. Nuclear power plants take a long time to build and are incredibly expensive—and are notorious for going massively over budget. Canada has subsidized the nuclear power industry to the tune of $20 billion over the past fifty years. In the U.S., "subsidies to the nuclear fuel cycle have often exceeded the value of the power produced," reaching into hundreds of billions of dollars over fifty years, according to a 2011 report by the Union of Concerned Scientists.[85] Just think of what could have been done by putting that kind of money into renewable energy.

Nuclear energy isn't even all that green when it comes to global warming. If you look at the life cycle of nuclear power plants, the technology produces greenhouse gases at every step, from energy-intensive uranium mining and transportation to constructing and decommissioning power plants. (Looking at the life cycle of energy technologies hasn't always been a common practice, but it's an important step that has allowed us to identify problems with energy sources that look attractive at first glance, such as corn-based biofuels.) The waste from uranium mining and nuclear power plants is a serious issue, especially considering that much of it is highly radioactive. Although we can recycle some waste from power production, we still haven't really figured out what to do with most of it. One method for large-scale storage is to bury it, but that's basically a policy of out of sight, out of mind. We don't yet know the full consequences. It's also expensive and the waste has to be transported over long distances where the probability of a mishap is very real.

Uranium is also a limited resource. The European Commission estimated in 2001 that global supplies of uranium might only last from about forty to seventy years with current usage rates. Prices have already skyrocketed as uranium becomes scarce. Maybe one of the biggest dangers is that by-products from nuclear power generation can be used to make terrifying weapons, and there are very real fears that widespread use of nuclear could put these products in the hands of terrorists and rogue states.

And although nuclear has a relatively good safety record compared to some other large-scale energy technologies, the consequences of an accident can be far worse. Although nuclear energy's ability to provide large-scale continuous power makes it tempting, advances in power-grid management and renewable energy technologies have given us better ways to deal with our energy needs. And, as sustainability writer Jeremy Lent noted in a 2016 *Huffington Post* article, because nuclear power requires "massive centralized investment along with extreme security," it tends to concentrate political power and decision-making in the hands of politicians and billionaires, unlike technologies such as solar power, which are more equitably available even to people with limited resources.[86]

Some argue, however, that problems such as radioactive waste, meltdown risks, weapons proliferation, and uranium scarcity can be overcome with safer nuclear technologies.

Even eminent climate scientists like James Hansen claim we can't avoid nuclear if we want to reduce greenhouse gas emissions. Hansen, a former NASA scientist, with Ken Caldeira of the Carnegie Institution for Science, Kerry Emanuel of the Massachusetts Institute of Technology, and Tom Wigley of Australia's University of Adelaide, wrote an open letter in 2013 stating, "the time has come for those who take the threat of global warming seriously to embrace the development and deployment of safer nuclear power systems."[87]

What are "safer nuclear power systems"? And are they the answer? Proposed technologies include smaller modular reactors, reactors that shut down automatically after an accident, and molten salt reactors.[88] Some would use fuels and coolants deemed safer. (Industry proponents argue the low incidence of nuclear accidents means current technology is safe enough. But the costs and consequences of an accident, as well as problems such as containing highly radioactive wastes, provide strong arguments against building new reactors with current technology.)

One idea is to use thorium instead of uranium for reactor fuel.[89] Thorium is more abundant than uranium. Unlike uranium, it's not fissile; that is, it can't be split to create a nuclear chain reaction, so it must be bred through nuclear reactors to produce fissile uranium. Thorium-fuelled reactors produce less waste, and although some trace elements in spent uranium fuels remain radioactive for many thousands of years, levels in spent thorium fuels drop off much faster. China and Canada are working on a modified Canadian design that includes thorium along with recycled uranium fuel. With the right type of reactor, such as this design or the integral fast reactor, meltdown risks are reduced or eliminated.

Thorium can be employed in a variety of reactor types, some of which currently use uranium—including heavy water reactors like Canada's CANDU. But some experts say new technologies, such as molten salt reactors, including liquid fluoride thorium reactors (LFTRS), are much safer and more efficient than today's conventional reactors.

So why aren't we using them?

Although they may be better than today's reactors, LFTRs still produce radioactive and corrosive materials, they can be used to produce weapons, and we don't know enough about the impacts of using fluoride salts.[90] Fluoride will contain a nuclear reaction, but it can be highly toxic, and deadly as fluorine gas. And although the technology's been around since the 1950s, it hasn't been proven on a commercial scale.[91] Countries including the U.S., China, France, and Russia are pursuing it, but in 2010 the U.K.'s National Nuclear Laboratory reported that thorium claims are "overstated."[92]

It will also take a lot of time and money to get a large number of reactors on stream—some say from thirty to fifty years. Given the urgent challenge of global warming, we don't have that much time. Many argue that if renewables received the same level of government subsidies as the nuclear industry, we'd be ahead at lower costs. Thorium essentially just adds another fuel option to the nuclear mix and isn't a significant departure from conventional nuclear. All nuclear power remains expensive, unwieldy, and difficult to integrate with intermittent renewables—and carries risks for weapons proliferation.

If the choice is between keeping nuclear power facilities running or shutting them down and replacing them with coal-fired power plants, the nuclear option is best for the climate—with thorough inspections to ensure they are operating safely. But, for now, investing in renewable energy and smart-grid technologies is a faster, more cost effective, and safer option than building new nuclear facilities, regardless of type.

That doesn't mean we should curtail research into nuclear and other options, including thorium's potential to improve the safety and efficiency of nuclear facilities and technologies that use current waste as fuels. But we must also build on the momentum of renewable energy development, which has been spurred by its safety, declining costs, and proven effectiveness.

## Natural Gas: Bridge to the Future or Dead-End Path?

NATURAL GAS IS often touted as a way to move away from burning fuels such as coal and oil. But can a fossil fuel really help us avoid the harmful effects of other fossil fuels? It's a question that comes up often as natural gas is eyed as a cleaner alternative or as a "transition" or "bridge" fuel that will help us shift from the "dirtier" alternatives. Burning coal and oil causes pollution and emits greenhouse gases that drive climate change. Exploring and drilling for oil and mining coal also come with numerous environmental impacts—especially as easily accessible oil runs out and we have to rely on deepwater drilling and oil sands. Natural gas burns cleaner than oil and coal, and it emits less carbon dioxide for the amount of energy it produces. This has led industry and governments to argue for an increase in natural gas production.

Russia, the United States, and Canada are the world's largest natural gas producers. Although overall production has been declining in Canada, sources and methods for exploiting "unconventional" natural gas reserves, such as shale gas, have led industry and government officials to argue that gas could play a role as a "bridging" fuel to kick-start near-term reductions in the greenhouse gas emissions responsible for climate change. The idea that natural gas could be a bridging fuel was one the industry came up with itself back in the early 1980s. It's not that simple, though, especially when we consider the impacts of unconventional natural gas, along with extraction methods such as hydraulic fracturing, or fracking. A 2011 report by the David Suzuki Foundation and Pembina Institute, *Is Natural Gas a Climate Change Solution for Canada?*, examined the key issues around natural gas and reaches surprising conclusions.[93]

Extracting gas from shale deposits, for example, requires up to one hundred times the number of well pads to get the same amount of gas as conventional sources. Imagine the disruption in farm or cottage country of one well pad (comprising multiple wells) roughly every 2.5 square kilometres. Each well pad occupies an area of about one hectare and also requires access roads and pipeline infrastructure. Leaking natural gas, which is mostly the powerful greenhouse gas methane, also contributes to global warming. According to an article in *Nature*, scientists from the U.S. National Oceanic and Atmospheric Administration and the University of Colorado Boulder found methane leaks amounted to between 4 and 9 percent of total production at two gas fields in the U.S.

Burning natural gas and the industrial activity required to extract, liquefy, and transport it also contribute to increased greenhouse gas emissions. According to the Pembina Institute, if only five of twelve proposed liquefied natural gas terminals were built on the B.C. coast, they could spew 69 million tons of carbon a year into the atmosphere—exceeding the amount produced by the Alberta oil sands and equal to all of B.C.'s greenhouse gas emissions in 2010. Discharges of particulate matter and volatile organic compounds would also be significant new sources of pollution. The U.S. National Center for Atmospheric Research has concluded that switching to natural gas won't do much to help solve the climate crisis.

Another problem is that fracking has become a common method to extract natural gas. Although it has been used to

extract gas since the late 1940s, producers only began combining it with horizontal drilling to exploit unconventional
gas resources since the turn of this century. With this process,
water, sand, and chemicals are pumped at high pressure into
rock formations deep in the earth to fracture the rock, allowing
the gas to escape and flow into the wells. In many jurisdictions,
companies are not required to disclose the chemicals they use
for fracking. Hydraulic fracturing requires massive amounts
of water. Disposing of the toxic wastewater, as well as acciden-
tal spills, can contaminate drinking water and harm human
health.[94] Gas leakage also leads to problems, even causing tap
water to become flammable! In some cases, flaming tap water
is the result of methane leaks from fracking. As noted earlier,
methane is a far more potent greenhouse gas than carbon
dioxide!

The enormous amounts of water used for fracking present a serious problem. A 2014 report, *Hydraulic Fracturing &
Water Stress*, shows the severity of the problem.[95] The study
by U.S. nonprofit Ceres found that hydraulic fracturing is
using enormous amounts of water in areas that can scarcely
afford it. The report noted that close to half the oil and gas
wells recently fracked in the U.S. "are in regions with high or
extremely high water stress," and more than 55 percent are in
areas experiencing drought. In Colorado and California, almost
all wells—97 and 96 percent, respectively—are in regions
with high or extremely high water stress, meaning more than
80 percent of available surface and groundwater has already
been allocated for municipalities, industry, and agriculture.
A quarter of Alberta wells are in areas with medium to high
water stress.

Drought and fracking have already caused some small communities in Texas to run out of water altogether, and parts of

California are headed for the same fate. As we continue to extract and burn ever-greater amounts of oil, gas, and coal, climate change is getting worse, which will likely lead to more droughts in some areas and flooding in others. California's drought may be the worst in five hundred years, according to B. Lynn Ingram, an earth and planetary sciences professor at the University of California, Berkeley.[96] The drought is causing a shortage of water for drinking and agriculture, and for salmon and other fish that spawn in streams and rivers. With no rain to scrub the air, pollution in the Los Angeles area has returned to dangerous levels of decades past.

Because of lack of information from industry and inconsistencies in water volume reporting, Ceres's western Canada data analysis "represents a very small proportion of the overall activity taking place." Researchers determined, though, that Alberta fracking operations have started using more "brackish/saline" groundwater instead of freshwater. The report cautions that this practice needs more study "given the potential for brackish water to be used in the future for drinking water" and the fact that withdrawing salty groundwater "can also adversely impact interconnected freshwater resources."

Although B.C. fracking operations are now mainly in low water stress regions, reduced precipitation and snowpack, low river levels, and even drought conditions in some areas— probably because of climate change—raise concerns about the government's plan to rapidly expand the industry. The report cites a "lack of regulation around groundwater withdrawals" and cumulative impacts on First Nations lands as issues with current fracking.

Ceres's study only looks at fracking impacts on freshwater supplies, and offers recommendations to reduce those, including recycling water, using brackish or wastewater, strengthening regulations, and finding better ways to dispose of fracking wastewater. But the drilling method comes with other environmental problems, from groundwater contamination to massive ecosystem and habitat disruption all done in the name of short-term gain.

As if the water-related problems weren't enough, fracking is also known to cause earthquakes.[97] The B.C. Oil and Gas Commission has reported that numerous quakes and tremors in that province were caused by fracking.[98] Studies have found quakes are common in many places where that natural gas extraction process is employed. It's not unexpected that shooting massive amounts of water, sand, and chemicals at high pressure into the earth to shatter shale and release natural gas might shake things up. But earthquakes and water use and contamination aren't the worst problem with fracking.

The biggest issue is that it's just one more way to continue our destructive addiction to fossil fuels. As easily accessible oil, gas, and coal reserves become depleted, corporations have increasingly looked to unconventional sources, such as those in the oil sands or under deep water, or embedded in underground shale deposits. And so we end up with catastrophes such as the spill—and deaths of eleven workers—from the *Deepwater Horizon* blowout in the Gulf of Mexico in 2010. We turn a blind eye to the massive environmental devastation of the oil sands, including contamination of water, land, and air; destruction of the boreal forest; endangerment of animals such as caribou; and impacts on human health. We blast the tops off of mountains to get coal. We figure depleted water supplies, a few earthquakes, and poisoned water are the price we have to pay to maintain our fossil-fuelled way of life.

As Bill McKibben points out, it didn't have to be this way. "We could, as a civilization, have taken that dwindling supply and rising price as a signal to convert to sun, wind, and other noncarbon forms of energy," he wrote in the *New York Review of Books* in 2012, adding that "it would have made eminent sense, most of all because it would have aided in the fight against global warming, the most difficult challenge the planet faces."[99]

In the short term, we must realize that we have better ways to create jobs and build the economy than holding an "everything must go" sale on our precious resources. In the longer term, we must rethink our outdated economic systems, which were devised for times when resources were plentiful and infrastructure was scarce. Our highest priorities must be the air we breathe, the water we drink, the soil that provides food, and the biodiversity that keeps us alive and healthy.

The non-climate environmental impacts of gas extraction alone are enough to give us pause, but natural gas is not even a good way to fight climate change. More than anything, continued and increasing investment in natural gas extraction and infrastructure will slow investment in, and transition to, renewable energy. Would companies that build gas-fired power plants be willing to shut them down, or pay the high cost of capturing and storing carbon, as the world gets serious about the need to dramatically reduce greenhouse gas emissions? Just as fossil fuels from conventional sources are finite and becoming depleted, those from difficult sources will also run out. If we put all our energy and resources into continued fossil fuel extraction, we will have lost an opportunity to invest in renewable energy.

The real solutions to climate change lie with conservation and renewable energy, such as solar, wind, tidal, and geothermal power. But because natural gas will be with us for the foreseeable future, we must do all we can to clean up practices associated with it. That means requiring industry to disclose the chemicals used in fracking and implementing strict regulation, monitoring, and environmental assessment processes.

If we want to address global warming, along with the other environmental problems associated with our continued rush to burn our precious fossil fuels as quickly as possible, we must learn to use our resources more wisely, kick our addiction, and quickly start turning to sources of energy that have fewer negative impacts. Using one fossil fuel to get off others isn't the way to go about it.

## Can We Geoengineer Our Way out of This Mess?

BECAUSE WE'VE ALREADY pumped so many greenhouse gases into the atmosphere and triggered feedback cycles, and because we don't have the means to stop using all fossil fuels right away, we can't stop global warming immediately. We can only hope to slow it down and eventually bring it under control. That's led many people to propose geoengineering to reverse some of the more disastrous effects of burning fossil fuels. Is this a viable solution?

First, to be clear, by "geoengineering," we do not mean "chemtrails." Those who believe in chemtrails claim governments around the world are in cahoots with secret organizations to seed the atmosphere with chemicals and materials—aluminum salts, barium crystals, biological agents, polymer fibers, etc.—for a range of nefarious purposes. These include controlling weather for military purposes, poisoning people for population or mind control, and supporting secret weapons programs based on the High Frequency Active Auroral Research Program, or HAARP, an ionic research facility and program based in Alaska that was shut down in 2014.

Scientists have tested and used cloud and atmospheric seeding for weather modification and considered them as ways to slow global warming. With so many unknowns and possible unintended consequences, these practices have the potential to cause harm. But the chemtrails conspiracy theory is much broader, positing that military and commercial airlines are involved in constant massive daily spraying that is harming the physical and mental health of citizens worldwide.

I don't have space to get into the absurdities of belief in a plot that would require worldwide collusion between governments, scientists, and airline company executives and pilots to amass and spray unimaginable amounts of chemicals from altitudes of nine thousand metres or more. I'm a scientist, so I look at credible science—and there is none for the existence of chemtrails. What people are claiming to be chemtrails are just condensation trails, formed when hot, humid air from jet exhaust mixes with colder low-vapour-pressure air. This, of course, comes with its own environmental problems.

It's interesting that many people who believe in a scientifically unsound idea like chemtrails dismiss the scientifically proven theory of anthropogenic global warming. It may be in part because both climate change denial and belief in chemtrails are often conspiracy-based.

Many deniers see climate change as a massive plot or hoax perpetrated by the world's scientists and scientific institutions, governments, the UN, environmentalists, and sinister forces to create a socialist world government or something. People who subscribe to unbelievable conspiracy theories may feel help-less, so they see themselves as victims of powerful forces—or as heroes standing up to those forces. Whether it's to deny real problems or promulgate imaginary ones, it helps reinforce a world view that is distrustful of governments, media, scientists, and shadowy cabals variously referred to as banksters, global elites, the Illuminati, or the New World Order.

The only connection between chemtrails and geoengi-neering as applied to climate change is that, if it were valid, the chemtrails idea would be an example of geoengineering. Geoengineering as it relates to global warming solutions has been tried and is being proposed for wider application, but it's not as secretive, widespread, or nefarious as supposed chemtrails.

Although it has been employed in limited applications, geo-engineering to combat climate change is largely untested. But because we've stalled so long on reducing carbon emissions, and still aren't doing enough, we may have to consider it. What will that mean?

As it relates to climate change, geoengineering falls into two categories: solar radiation management and carbon cap-ture and storage. The former involves reflecting solar radiation back into space. The latter is aimed at removing carbon dioxide from the atmosphere and storing it.

Solar radiation management includes schemes such as releasing sulphur aerosols into the atmosphere to scatter sun-light and reduce radiation, creating or whitening clouds by spraying seawater or other materials into the air, and even installing giant reflectors in space. These methods don't affect $CO_2$ levels and so don't address issues such as ocean acidifica-tion, but they offer possible quick fixes to reduce warming.

An example of possible carbon removal is fertilizing oceans with iron. Iron stimulates growth of small algae called phytoplankton, which remove carbon dioxide from the sea and release oxygen through photosynthesis. This allows the oceans to absorb additional $CO_2$ from the atmosphere. The idea is that the plankton die and sink to the ocean floor, and become buried under other materials, storing the carbon within them. This method was tried in my part of the world, coastal B.C., in an incident of "rogue" geoengineering. An American businessman working with the Haida village of Old Massett dumped more than 100 metric tons of iron sulphate into the ocean in 2012 for a salmon restoration and carbon-reduction project. Although it may have had an effect on salmon populations, many researchers are skeptical about its carbon-storing benefits. According to *Scientific American*, "As other iron fertilization experiments have shown, it is relatively easy to get plankton to bloom, but it is harder for that bloom to sink to the bottom of the ocean, where it takes $CO_2$ with it. Instead, as suggested by the trickle-up theory of salmon restoration, the plankton tends to get eaten by tiny animals, which are then eaten by larger animals until, ultimately, all or most of the $CO_2$ sucked up by the tiny plants during their photosynthetic life spans finds its way back to the atmosphere in relatively short order."[100]

A 2009 opinion article in *Nature* by a group of oceanographers cautioned against iron fertilization, because of possible unintended consequences and ineffectiveness: "Given that the efficacy and indirect effects of ocean fertilization for geoengineering cannot be tested directly without altering the ocean on unprecedented scales, we must resort to using global-ecosystem models to predict its promise and pitfalls. Many modelling analyses have shown that iron fertilization cannot reduce atmospheric $CO_2$ enough to significantly alter the course of climate change."[101] And in a paper in the journal *Proceedings of the National Academy of Sciences of the USA*, scientists reported that the process can cause blooming of plants that produce deadly neurotoxins.[102] Oops.

According to a 2015 report by the Australia-based Global Carbon Capture and Storage Institute, only fifteen large-scale CCS projects were in operation worldwide, with "capacity to capture up to 28 million tonnes of $CO_2$ per year," with another seven or so slated to come online by 2017 or 2018.[103] The projects include SaskPower's Boundary Dam Unit 3 coal-fired carbon capture and storage facility in Saskatchewan. Touted by the Saskatchewan government as a great success, the taxpayer subsidized $1.1 billion project "has been plagued by multiple shutdowns, has fallen way short of its emissions targets, and faces an unresolved problem with its core technology. The costs, too, have soared, requiring tens of millions of dollars in new equipment and repairs," according to the *New York Times* in 2016.[104]

The governments of Canada, Saskatchewan, and Alberta have spent billions on plans and projects to trap $CO_2$ released by burning fossil fuels and pump it into the ground—but this method has yet to be perfected. It was developed when people in the oil industry found that as they drained oil from wells, they could pump $CO_2$ back in to mix with and recover more oil, a process called enhanced oil recovery. And the $CO_2$ appeared to stay in the ground. But we have no idea what happens to this gas. Does it form a bubble under a big rock? Is it chemically bonded to its surrounding matrix? How long will it stay down there? We don't know. We air-breathing terrestrial beings seem to have the attitude of "out of sight, out of mind," so we dump our garbage into the oceans or the ground or the atmosphere, as if that were a solution.

Until a few years ago, scientists assumed no life existed below bedrock, but miners kept reporting that bits drilled far deeper into the ground came back contaminated. Researchers later discovered bizarre forms of life three kilometres below the surface. The organisms are bacteria, which in some cases are embedded in rock, eking out an existence scrounging for water, energy, and nutrition. Some are thought to divide only once in a thousand years! When these organisms are brought to the surface, their DNA is unlike anything we know about bacteria aboveground. Biologists have had to invent whole new phyla to describe them. The layer of life on Earth's surface is very thin, but these single-celled organisms go down miles. Now, scientists believe that protoplasm living underground are more abundant than all of the elephants, trees, whales, fish, and other life above! We have no idea how important these organisms are to the subsurface web of life. Do they play a role in movement of water and nutrients, of energy from the magma? We have no idea.

Another problem is that when carbon captured from CCS projects is used to recover oil from near-depleted wells, the carbon emitted from burning the recovered oil can exceed the amount stored. And rather than encouraging renewable energy development, CCS is a way to continue exploiting fossil fuels.

Carbon capture and storage may be worth studying, but the technology's potential should not be used as an excuse for the oil and coal industries to avoid reducing their emissions and investing in renewable energy. After all, we know that energy conservation and renewable energy will yield the immediate effects of a cleaner environment. We don't know what carbon capture and storage will cost, when it will be commercially viable, or what it will do, other than perhaps to give us a way to keep relying on finite and polluting sources of energy.

Many geoengineering schemes are controversial and have shown mixed results in tests, and the danger of unintended consequences is real, including further catastrophic, irreversible damage to the climate system. One major drawback with geoengineering is the mistaken idea that it can be a substitute for reducing greenhouse gas emissions that cause climate change. That many geoengineering projects are fraught with danger and would not resolve the problem quickly enough or even effectively—and would do little or nothing to resolve other fossil fuel problems such as pollution—makes this a critical concern.

There's also the matter of who would decide what methods to apply and when and where. A U.K. Royal Society study concluded that geoengineering "should only be considered as part of a wider package of options for addressing climate change," and carbon dioxide reduction methods should be preferred over more unpredictable solar radiation management.[105] Scientists at the Berlin Social Science Research Center suggest creating "a new international climate engineering agency... to coordinate countries' efforts and manage research funding."[106] Because some geoengineering is probably unavoidable, that's a good idea. But rather than rationalizing our continued use of fossil fuels in the false belief that technology will enable us to carry on with our destructive ways, we really need governments, scientists, and industry to start taking climate change and greenhouse gas emissions seriously. We can't just engineer our way out of the problem.

As sustainability writer Jeremy Lent wrote in a 2016 *Huffington Post* article, "Geoengineering proposals are based on the notion of the earth as a massive piece of machinery to be engineered for human benefit. Not only are these approaches morally repugnant for anyone who sees Nature as having intrinsic worth, they are also fraught with massive risk, since the earth's systems are in fact not machine-like, but the result of complex, nonlinear relationships that are inherently unpredictable."[107] Lent also argued that geoengineering concentrates political and decision-making power at the top: "Developed by small, elite groups of technical experts, usually funded by corporate interests, they frequently envisage forcing a global experiment on the entire world, regardless of whether there is a consensus supporting such an approach."

As Naomi Klein notes in *This Changes Everything*, geoengineering on the scale needed to put a dent in global warming would be incredibly costly with lots of uncertainty and possible unintended consequences. Better, she argued, to focus on the solutions we know will work: "After all, if the danger of climate change is sufficiently grave and imminent for governments to be considering science-fiction solutions, isn't it also grave and imminent enough for them to consider just plain science-based solutions?"[108]

Chapter 7

# INSTITUTIONAL SOLUTIONS

MANY PEOPLE AROUND the world have accepted the reality of climate change and the need to conduct our lives differently if humanity is to have a bright future. That means people have started making changes in their own lives—eating less meat and animal products or avoiding them altogether, driving less, and conserving energy, among other things. Entrepreneurs have taken advantage of the growing demand for clean energy and energy-efficient products, developing and marketing goods that fulfill human wants and needs without destroying the environment or the climate. Researchers are looking into ways to supply energy to growing populations without burning through finite, polluting, climate-altering fossil fuels. Others are testing various agricultural production and distribution methods to see how we can best feed 7 billion or more people with minimal damage to the planetary life-support systems that keep us alive and healthy.

With so many people living under so many different conditions and types of government, resolving a massive global problem like climate change also requires changes that can only come about with global agreements and national and international regulations. It may seem difficult, but we've seen how rapidly large global trade deals have altered the way the world does business in a relatively short time period. We've also seen how regulations that account for pollution and environmental destruction, and make polluters pay, can discourage damaging practices and encourage better ways of doing things. The slow pace and ongoing failures of global climate negotiations can be discouraging, but the 2015 Paris Agreement, though inadequate, was a breakthrough. It at least signalled a new way of looking at and talking about the climate crisis, and it calls for continued improvement in global efforts to reduce greenhouse gas emissions.

Bringing about the massive changes required to prevent catastrophic global warming will take a lot of cooperation. That's not easy but it's not impossible. It will also take rules and regulation, which despite what some free-market fundamentalists will tell you, shouldn't be seen as a bad thing. Whether or not we can resolve the climate crisis under existing systems and with current ways of thinking is a big question. But humans are capable of evolving and adapting to changing circumstances. In this chapter, we'll discuss the governance and agreements that are essential to making the necessary changes.

## Global Agreements and the New Language of Climate Change

IF NOTHING ELSE, the Paris Agreement to limit greenhouse gas emissions and an earlier commitment by G7 countries to end fossil fuel use for energy by 2100 signal a shift in the way we talk and think about global warming. The G7 agreement goes beyond previous agreements—which were about reducing carbon emissions from burning coal, oil, and gas—to envisioning a fossil fuel–free future.[1]

But it's easy to be cynical about these and other international commitments and agreements, for a number of reasons: the long time frame for the G7 pledge, especially, means none of the politicians involved in the commitment will even be alive, let alone held accountable, for meeting the target in 2100; Canada and Japan watered down Germany's proposal to end fossil fuel energy by 2050; and many governments, including Canada's, haven't met even their current weak commitments. But in calling for deep emissions cuts by 2050, and an end to fossil fuel energy by 2100—"decarbonization"—the nonbinding pledge at least shows governments recognize the need to confront climate change.

Because of the oil sands, Canada has a significant role to play, and could show it takes the commitment seriously by heeding the advice of one hundred scientists (including twelve Royal Society of Canada fellows, twenty-two U.S. National Academy of Sciences members, five Order of Canada recipients, and a Nobel Prize winner, from a range of disciplines) who released a statement with ten reasons why "no new oil sands or related infrastructure projects should proceed unless consistent with an implemented plan to rapidly reduce carbon pollution, safeguard biodiversity, protect human health, and respect treaty rights."[2]

According to Simon Fraser University energy economist and statement coauthor Mark Jaccard, "Leading independent researchers show that significant expansion of the oil sands and similar unconventional oil sources is inconsistent with efforts to avoid potentially dangerous climate change."

Another author, Northern Arizona University ecologist Tom Sisk, said it's not just about climate: "Oil sands development is industrializing and degrading some of the wildest regions of the planet, contaminating its rivers, and transforming a landscape that stores huge amounts of carbon into one that releases it."

The reasons for a moratorium include: oil sands expansion and investment are incompatible with climate protection and are slowing the shift to clean energy, monitoring and enforcement are inadequate, landscape is being contaminated and reclamation is insufficient, First Nations treaties are being violated, affordable alternatives are available, cumulative impacts have been ignored, and Canadians are demanding solutions.

Of course, it will take more than signing a nonbinding pledge and slowing or halting oil sands expansion to avert the worst consequences of climate change. In a 2014 article in Nature, eight scientists who signed the moratorium statement, including Jaccard, argued Canada and the U.S. must stop treating "oil-sands production, transportation, climate and environmental policies as separate issues, assessing each new proposal in isolation. A more coherent approach, one that evaluates all oil-sands projects in the context of broader, integrated energy and climate strategies, is sorely needed."[3]

As part of a coordinated strategy, they proposed putting a price on carbon, through a carbon tax or cap and trade, to "ensure that the full social costs of carbon combustion are incorporated into investment decisions about energy and infrastructure." Carbon pricing is widely accepted as an effective way to discourage fossil fuel use and encourage clean energy development.

Moving toward zero carbon emissions—in a much shorter timeline than agreed upon by Canada, France, Germany, Italy, Japan, the United Kingdom, and the United States—is absolutely necessary, and not just for the climate. Eliminating fossil fuel energy will cut dangerous pollution, create new economic opportunities, and ensure resources are available for wiser applications.

The words of scientists, government leaders, and other experts—even religious leaders such as Pope Francis and the Dalai Lama—make it clear that it's time to turn the page on this destructive and relatively recent chapter in our history. Now we must ensure our leaders strengthen and act on their commitments.

Despite the warranted cynicism about the ability of government agreements to accomplish the necessary reductions, we've seen that international leadership based on sound science can lead to great results. For proof, we need only to look up. The ozone layer is no longer shrinking. Starting in the 1970s, scientists observed a connection between our use of CFCs, and a weakening of the ozone layer in the stratosphere. High above Earth, ultraviolet light breaks chlorine off the CFC molecule, and chlorine is a potent scavenger of ozone. Stratospheric ozone absorbs ultraviolet radiation, protecting us from the sun's rays like a giant pair of sunglasses.

CFCs were once used in products ranging from aerosol spray cans to refrigerators. As more of the chemicals were dumped into the air, they began to destroy the ozone layer, creating the potential for dramatic increases in skin cancers and damage to the phytoplankton that form the base of life. In September 1987, world leaders signed the Montreal Protocol on Substances that Deplete the Ozone Layer. A 2010 report written and reviewed by three hundred scientists from around the world concluded that phasing out production and consumption of ozone-depleting substances under the Montreal Protocol "has protected the stratospheric ozone layer from much higher levels of depletion."[4]

It's not a complete turnaround, but it is good news. The scientists found that global ozone and ozone in the Arctic and Antarctic regions are no longer decreasing, but they are not yet increasing either. They also wrote, "the ozone layer outside the Polar regions is projected to recover to its pre-1980 levels some time before the middle of this century."

UN Environment Programme executive director Achim Steiner noted that, without the agreement, atmospheric levels of ozone-depleting substances could have increased tenfold, leading to "up to 20 million more cases of skin cancer and 130 million more cases of eye cataracts, not to speak of damage to human immune systems, wildlife and agriculture."[5] Interestingly, the scientists and world leaders who worked to protect us from ozone depletion faced many of the same pressures that those working to protect us from climate change now encounter. CFC manufacturers claimed that the science on the dangers of CFCs was "rubbish" and that phasing out CFCs would cost trillions of dollars and would destroy the industry. As Naomi Oreskes and Erik Conway write in *Merchants of Doubt*, many of the same "experts" show up in the campaigns industry has waged against the science regarding the impacts of tobacco, CFCs, acid rain, and climate change.[6]

If we can succeed in tackling the ozone problem, despite attacks from industry, why is it so difficult to resolve an even greater threat to life on the planet—climate change? One of the scientists who won a Nobel Prize for chemistry in 1995 for his work on the ozone layer has an explanation. Sherwood Rowland said in a 2010 interview that "arguing which propellant to use was rather trivial to society. One could replace CFCs and still use existing technology. This is quite different from having fossil fuels as our primary energy source for the whole world."[7]

In other words, the stakes are higher—for industry and society. In many cases, CFCs could be replaced by something as simple and nonpolluting as compressed air. And despite the claims of chemical manufacturers, phasing out CFCs did not bankrupt the industry, because these chemicals were only one product among many that the companies produced. Although some energy companies are working on clean energy technology, their massive profits come mainly from exploiting ever-dwindling supplies of fossil fuels. And pretty much everyone in the world relies on fossil fuels to some extent. The good news is that in the past two years total worldwide investments in renewable electricity generation were greater than total investments in fossil fuel–based electrical capacity.

Coming to global agreement that puts the brakes on the most profitable and entrenched industry in the history of humanity is a challenge, to be sure—something that is clear from the decades of failed climate negotiations and even the inadequate Paris Agreement. But governments have much less difficulty in negotiating, implementing, and upholding trade agreements. There's no reason why, if climate change is recognized as the serious threat to humanity that it is, they can't agree on measures necessary to ensure that humans actually have a future on this planet—especially if the people they are supposed to represent demand that they do so. There's no reason protection for the climate and environment can't be incorporated into trade agreements, either, though many current agreements put corporate interests ahead of environmental protection.

The solutions exist, but it will take a lot of effort and political will to make the shift. If we do it right, it will have enormous benefits for human health and economies. But don't expect the most profitable industry in the history of the universe to get onboard anytime soon. It's up to all of us to demand change. The Montreal Protocol and other global agreements show that governments of all types can find common ground on a range of beneficial issues, but we must all listen to reason rather than the claims of those who put profits before people, and we must demand that governments act in the interests of people and not just corporations.

## Paying the Price to Pollute

IT'S WRONG THAT people and corporations are allowed to spew pollutants and emissions into the atmosphere without costs or consequences. Carbon pricing, through carbon taxes or cap-and-trade systems, is designed to address that oversight and to encourage householders and corporations to pollute less. As former CIBC World Markets chief economist Jeff Rubin and I have argued, "Carbon pollution is a classic example of a market failure where polluters aren't required to pay for the damage they create. In order that they do, government has to set in place the right policy framework."[8]

A carbon tax is simply a fee placed on carbon dioxide emissions, mainly from burning fossil fuels. This makes it more expensive to use coal, oil, and gas and thus provides an incentive to reduce their use through conservation and using cleaner energy sources. That raises concerns that lower-income people could be disproportionately affected, but jurisdictions such as British Columbia offer tax credits to offset those effects. The B.C. carbon tax is also revenue neutral; that is, other tax reductions mean that overall taxes paid by citizens are no higher than before the carbon tax was introduced.

With cap and trade, an overall limit, or cap, is set on the amount of greenhouse gas emissions a jurisdiction (province, state, or country) can emit. It then tells polluters, such as heavy industry and electricity generators, how many tons of carbon each can release. For every ton, polluters need a permit or "allowance." So, if a company's annual limit were 25,000 tons, it would require 25,000 allowances. If a company exceeds its limit, it can purchase additional allowances from another firm that, because of its greater efficiency, has more allowances than it needs. This is the "trade" part of the equation.

Although an individual company can exceed its greenhouse gas limit by purchasing credits, the jurisdiction as a whole can't. The overall limit is reduced every year, so if the law is followed, cap and trade guarantees annual emissions reductions. The declining cap is the system's great strength and the way it protects the environment. By balancing out the emissions cuts, overall cuts to emissions meet the targets set even if some companies emit more than the cap allows.

Quebec, California, and the European Union have already adopted cap and trade, and Ontario will join Quebec and California's system in January 2017.

How effective is cap and trade? Although the answer isn't straightforward, there's evidence it played a key role in reducing acid rain in the United States. The 1990 Clean Air Act allowed power plants to buy and sell the right to emit sulphur dioxide. Since then, U.S. sulphur dioxide concentrations have gone down by more than 75 percent.[9] As economist Paul Krugman wrote in the *New York Times* in 2010, "Acid rain did not disappear as a problem, but it was significantly mitigated."[10]

Despite this and other successes, some experts are skeptical, arguing that cap and trade amounts to little more than a cash grab by government, a tax in everything but name. Others say it's a mistake to expect climate change to be addressed through markets, when the problem actually requires changing our entire approach to economics, with a commitment to a steady-state economy and an end to the commodification of nature.[11] Others have pointed out that there's too much room for manipulation or outright corruption, and that cap and trade sometimes means vulnerable people are put at risk, as polluting industries are able to buy allowances and keep polluting, often near disadvantaged communities.

Some experts have also noted that the emissions reductions it brings are often modest. A 2015 paper in *Canadian Public Policy* claimed Quebec's system "is still too weak to meaningfully address the environmental imperatives as outlined in the Intergovernmental Panel on Climate Change's 2014 *Fifth Assessment Synthesis Report*, in which fully eliminating carbon emissions is the benchmark for long-term policy goals."[12] From 2013 to 2014, California's allowance cap went from 162.8 to 159.7 megatons, a drop of less than 2 percent.[13]

Ontario's proposed legislation indicates its program will have some great strengths and a number of shortcomings. It will probably have wide coverage, applying limits on most of the province's emissions, including those from transportation fuels. (California's system did not initially include these fuels.)

Ontario is expected to reduce emissions by more than 4 percent a year—about twice the initial rate of California—and generate $1.9 billion annually from the plan. That money will be invested in green projects throughout the province with the goal of reducing carbon emissions even further.

To keep the bulk of fossil fuels in the ground—as scientific evidence says we must—we need a variety of strategies. Cap and trade helps reduce emissions and generates billions of dollars for other strategies to address climate change. It also embodies the polluter pays principle. But it's not enough on its own.

Carbon taxes have proven to be effective. British Columbia's carbon tax of $30 a metric ton has led to a decrease in motor fuel consumption, without a corresponding decline in the economy. Sweden imposed a carbon tax of US$133 a metric ton in 1991, and raised it to $168 in 2014. Emissions fell by more than 40 percent below 1990 levels between 1991 and 2008, and Sweden's economy grew by 44 percent.

Which is best, carbon taxes or cap and trade? If both approaches are well designed, the two options are similar and could be used in tandem. For example, if a federal minimum carbon price were set in Canada, with provinces and territories permitted to adopt their own systems, the federal carbon tax could be added to Quebec and Ontario's cap-and-trade system to ensure that the same combined price on carbon is applied throughout the country. The design of any carbon-pricing system will determine the environmental and economic effectiveness. For example, how strong is the economic incentive (i.e., the carbon price) to reduce emissions and switch to cleaner energy? To which emission sectors does the system apply? And how are the revenues used? Are they invested in green infrastructure or corresponding tax breaks?

The World Bank and others have argued that a $50-a-ton tax on carbon emissions applied throughout all developed countries could raise $450 billion a year, which could be used for clean energy development and other climate change solutions.

As Rubin and I have argued, rather than offering exemptions to strategically important vulnerable industries, carbon-pricing regulations could be built into trade policies. That means imported goods would be subject to the same carbon pricing as domestic goods, because "carbon pollution is an unfair trade subsidy that needs to be countervailed with a tariff."[14]

What's important is that the price on carbon pollution provides an incentive for everyone, from industry to households, to be part of the solution. Ultimately, the critical factor in reducing heat-trapping emissions is the strength of the economic signal. A stronger carbon price will kick-start more growth in clean, renewable energy and will encourage adoption of greener practices.

### CARBON OFFSETS: INDULGENCES OR REAL SOLUTIONS?

MOST CARBON OFFSETS are a voluntary form of carbon pricing, though "compliance" markets also exist for companies and governments to meet emissions-reduction targets. Some people compare voluntary carbon offsets to indulgences granted by the Church allowing sinners to avoid punishment for some transgressions. Others argue that offsets can be one of many legitimate tools used to tackle climate change, and that high-quality carbon offsets can result in real reductions in greenhouse gas emissions.

Carbon offsets are a popular way for individuals, businesses, and even governments to reduce their impact on the environment. The "voluntary" carbon market, made up of all these purchases of carbon offsets, increased in value globally from

$305 million in 2007 to $460 million in 2008, but dropped to $379 million in 2013—in part because many jurisdictions have put a price on carbon, through carbon taxes and/or cap and trade, meaning large emitters are required by law to offset or reduce emissions.[15]

A carbon offset is a credit for a reduction in greenhouse gas emissions generated by one project, such as a solar power installation, that can be used to cancel out the emissions from another source. Carbon offsets are typically measured in tons of $CO_2$ or their equivalent. Those who buy offsets are essentially investing in other projects that reduce emissions on their behalf, either because they are unable to do so themselves or because it is too expensive to make their own reductions. They're commonly used for air travellers, who pay a premium based on the amount of greenhouse gas emissions their travel would represent.

One thing to note is that not all carbon offsets are created equal. Because the market is largely unregulated, some offsets are unlikely to have any benefit for the climate. This is one reason why carbon offsets have gotten a bad rap.

What makes a good offset? Opinions vary on some of the finer points, but most experts agree that several conditions are necessary. Good offsets are additional; that is, they result in greenhouse gas reductions that wouldn't have otherwise occurred without the incentive of carbon offsets. For example, if a company is required by regulation to install technology to reduce emissions from its factory, the resulting emission reductions should not be sold as offsets.

A good carbon offset should also result in permanent reductions in greenhouse gas emissions. This is one reason why some organizations, including the David Suzuki Foundation, recommend against using tree planting to generate offsets. Although trees have many benefits for the environment, they make risky carbon offsets because they are susceptible to fire, logging, and insect infestation—any one of which can release the stored carbon back into the atmosphere and render the offset worthless. Permanent offsets are those that create ongoing reductions in emissions, such as offsets for projects like solar and wind installations that permanently reduce the need to burn fossil fuels for energy.

Good carbon offsets should also be verified by qualified auditors to ensure that the reductions have actually taken place.

Carbon offsets that are additional, permanent, and real can have a direct positive impact on the climate. And they can create some other important benefits. They provide money for much-needed renewable energy and energy-efficiency projects, which can help move society away from fossil fuels and toward a clean energy economy. Buying carbon offsets can also help to deal with emissions that aren't currently covered by government regulations, such as international air travel. And carbon offsets can also put a value on carbon, and help to educate businesses and consumers about the climate impact of their daily decisions, and where they should target their own reduction efforts.

Of course, people should do everything they can to reduce their greenhouse gas emissions, but when that isn't possible or feasible, buying high-quality offsets at least ensures that an equivalent amount of reductions is made elsewhere.

Carbon offsets alone won't solve climate change. We still need to find ways to make deep reductions in our own emissions. But the problem of climate change is so massive that it requires a whole range of solutions, and offsets can be part of that.

## Feed-In Tariffs Help Renewable Energy Grow

IN THE EARLY 1990s, Germany launched *Energiewende*, or energy revolution, a program "to combat climate change, avoid nuclear risks, improve energy security, and guarantee competitiveness and growth."[16] Renewable energy grew from 4 percent in 1990 to more than 27 percent in 2014, including a significant increase in citizen-owned power projects, according to energy think tank Agora Energiewende.[17]

Germany's greenhouse gas emissions dropped by 27 percent during that time. Its goal is to reduce emissions 40 percent from 1990 levels by 2020 and more than 80 percent by 2050. Polls show that 90 percent of Germans like the program—even though it means paying higher rates for electricity.

There's good reason for this widespread support. The primary technologies of wind and solar have become cost competitive with conventional energy sources. Variable renewable sources and flexibility options for conventional and renewable power generation are making baseload power obsolete—which means the system is geared to wind and solar rather than nuclear or coal. Germany's energy system is one of the most reliable in the world. And it's created jobs and revenue.[18]

*Energiewende* hasn't solved all of Germany's emissions and energy issues. Electricity rates are among Europe's highest, though they're expected to come down as more renewable energy becomes available, and efficient usage means "actual costs to households are comparable to countries with lower prices but higher consumption levels."[19] The country still gets more energy from coal than renewables, transportation and heating consume significant energy from conventional sources, and heavy industry makes Germany one of Western Europe's highest emitters.[20] Opposition from power utilities and coal companies, with consequent government compromises, has also slowed progress. But a range of initiatives and tools has put Germany on track to meeting its long-term climate commitments.

One tool Germany used to achieve its rapid progress was a feed-in tariff guaranteeing renewable energy producers— individuals, businesses, community organizations, and power companies—access to the grid and payment from power utilities for energy they put into the system. At first, the tariff wasn't enough to cover costs, but in 2000, Germany introduced a law that guaranteed feed-in tariffs for twenty years at prices high enough for producers to profit. As renewable energy costs drop and more is brought into the system, tariffs go down.

Feed-in tariffs are in place in many countries worldwide. Canada's Pembina Institute notes they're effective for several reasons. They "reward actual production" rather than just installation, they minimize development investment risks and "facilitate access to financing," and they encourage small, medium and large producers and "community and local ownership and engagement, minimizing opposition to projects."[21]

They also "encourage renewable power producers to use the most efficient technology, driving down costs by fostering industrial competition," and while they cause short-term electricity price hikes, those stabilize over time as costs and risks of conventional power generation and volatile fossil fuel markets are reduced. Income paid through tariffs "more than offsets any electricity price increases" for those who generate renewable energy. And if the full environmental and health damages of fossil fuels are considered, renewables are an even better bargain.

Many jurisdictions with feed-in tariffs have become leading exporters of renewable energy technology, creating local jobs and strengthening economies—with little or no government spending! Feed-in tariffs vary in rates and designs according to what types and scales of technologies governments want to encourage and where they want them located, which means they must be carefully designed.

Massive centralized power sources are not efficient and are quickly becoming outdated. Some power is lost when it has to be transmitted over long distances, and large sources usually keep operating even when power isn't required. Using smart grids and distributed renewable energy with demand-management systems allows energy to be dispatched where and when it's needed, most often over shorter distances, and a variety of power sources makes them more reliable, as power outages are less frequent.

Burning finite fuels in huge plants to generate electricity is outdated. Meeting global commitments to reduce greenhouse gas emissions and slow global warming requires a transition from fossil fuels to renewable energy. Feed-in tariffs are an effective way for governments to encourage that shift.

## New Economies for a Changing World

IN *THIS CHANGES EVERYTHING*, Naomi Klein argues that addressing climate change is not possible "without challenging the fundamental logic of deregulated capitalism" and that market mechanisms such as carbon pricing don't go far enough.[22] She may be right. But even within the current economic paradigm, we can no longer afford to accept the faulty logic of those who argue that addressing climate change is just too costly.

In failing to act on global warming, many leaders are putting jobs and economic prosperity at risk. Those who refuse to take climate change seriously are subjecting us to enormous economic risks and forgoing the numerous benefits that solutions would bring.

The price we will pay to fight climate change is a good investment in a healthy and prosperous future. Some of the costs include investments in public transit and renewable energy, in programs to reduce greenhouse gas emissions in other parts of the world, and in helping people cope with higher transportation and home-heating costs during the time of transition. Failing to make these investments will end up costing us a whole lot more.

As noted in the section about economic barriers in Chapter 3, both the World Bank and the 2014 U.S. nonpartisan report *Risky Business* argue that the economic benefits of addressing climate change far outweigh the risks."[23, 24] Both suggest carbon pricing as a way to address the climate crisis, with the World Bank arguing for "regulations, taxes, and incentives to stimulate a shift to clean transportation, improved industrial energy efficiency, and more energy efficient buildings and appliances."[25]

But efforts by organizations such as the World Bank to find economic solutions that don't disrupt our current free-market ideologies and systems are probably not enough. They illustrate that arguments pitting environment against economy are false and damaging, but they don't get us where we need to go fast enough. It's time for a twenty-first-century way of economic thinking that reflects the current realities, that takes into account resource scarcity and climate change and aims for greater equity.

As biological creatures, we depend on clean air, water, soil and energy, as well as biodiversity, for our well-being and survival. Surely protecting those fundamental needs should be our top priority and should dominate our thinking and the way we live. After all, we are animals and our biological dependence on the biosphere for our most basic needs should be obvious. The economy is a human invention, a tool that can be changed when it no longer suits our needs. The environment is the very air, water, land, and diversity of plant and animal life we cannot live without. Why not work to build a healthy, prosperous economy that protects those things?

Volumes of research conclude that ignoring climate change will be far more costly than addressing it, but economic systems that rely on constant growth and ever-increasing consumption for its own sake exacerbate the problem. The massive bills for cleaning up after events related to extreme weather, such as flooding and wildfires, are already rising every year. Climate change is also affecting water supplies and the world's ability to grow food, and is contributing to a growing number of refugees. According to the World Health Organization, close to 150 million people are already dying every year from causes related to global warming—and that doesn't include death and illness related to pollution from burning fossil fuels.[26]

A better economic vision would support the right of all people to live in a healthy environment, with access to clean air and water and healthy food. It would respect planetary boundaries and provide the moral imperative to decrease growing income disparities. Businesses would be required to pay for environmental damage they inflict, capital would be more widely distributed, and ideas, such as employee shareholder programs with ethically invested stocks, would be the norm. It would mean an end to working for the sake of working. Rejecting the idea of constant economic growth and the need to always have more would lead to societies where wealth is shared more equitably, where people get more of the things that really matter, including more time to spend enjoying life, being with friends and family, getting out in nature, volunteering in the community, and pursuing real interests. Changing the way we work—through shorter work hours and weeks and longer vacations, and increased options such as telecommuting—would also conserve energy, not to mention reduce stress, by cutting traffic congestion and the amount of energy needed to maintain workplaces.

This alternative economy would connect people to family, friends, and communities; focus on social capital investments over gross domestic product gains; and distribute wealth through taxes, social programs, and minimum guaranteed incomes. In *The Spirit Level: Why More Equal Societies Almost Always Do Better*, authors Richard Wilkinson and Kate Pickett write that developed countries with the greatest inequalities have higher rates of disease, mental illness, drug use, and a host of other social problems.[27] Reducing income gaps makes all of us healthier.

In many countries, we don't question our emphasis on constant growth. Economic and political systems are often aimed at short-term profits, ignoring how that affects long-term health and survival and leads to growing income inequality. Despite more than five decades of trying to fix our environmental challenges, forests are still threatened, deserts are spreading, and climate change is creating more frequent and intense storms, floods, forest fires, and droughts.

We also have to recognize that growing income inequality, which is a feature of current economic systems, threatens democracies and human health and survival. Why is it so difficult to imagine an economic vision aimed at caring for people and the planet?

Why are we rapidly exploiting finite resources and destroying precious natural systems for the sake of short-term profit and unsustainable economic growth? What will we do when oil runs out or becomes too difficult or expensive to extract if we haven't taken the time to reduce our demands for energy and shift to cleaner sources?

Why does our economic system place a higher value on disposable, often unnecessary, goods and services than on the things we really need to survive and be healthy? Sure, there's some contradiction in protesters carrying iPhones while railing against the consumer system. But this is not just about making personal changes and sacrifices; it's about questioning our place and the ways we live on this planet.

In less than a century, the human population has grown exponentially, from 1.5 billion to 7 billion—and it continues to grow. That's been matched by rapid growth in technology and products, resource exploitation, and knowledge. The pace and manner of development have led to a reliance on fossil fuels, to the extent that much of our infrastructure supports products such as cars and their fuels to keep the cycle of profits and wealth concentration going. Our current economic systems are relatively new—methods we've devised both to deal with the challenge of production and distribution for rapidly expanding populations and to exploit the opportunities.

It may seem like there's no hope for change, but we have to remember that most of these developments are recent, and that humans are capable of innovation, creativity, and foresight. Despite considerable opposition, most countries recognized at some point that abolishing slavery had goals that transcended economic considerations, such as enhancing human rights and dignity—and it didn't destroy the economies in the end, as supporters of slavery feared.

Let's be clear. Resolving a global problem like climate change will cost money in the short term. But the very survival of people—along with many plants and animals that we share this small planet with—could well be at stake if we don't. Former World Bank chief economist Nicholas Stern has estimated that to keep heat-trapping greenhouse gas emissions below levels that would cause catastrophic climate change would cost up to 2 percent of global GDP, but failure to act could cost from 5 to 20 percent of global GDP.[28] While we're addressing the immediate effects and costs, we have to start planning for economies that recognize we can no longer count on the abundance of resources we had when the current systems were developed.

Runaway climate change could have devastating impacts on our water and food supplies, could lead to waves of refugees escaping uninhabitable drought-stricken areas or vanishing islands, and could wreak havoc on the world's oceans and cause major extinctions of plants and animals. Some of this is already happening. Carbon taxes and incentives for clean energy development may not be enough to prevent that.

Some argue that it's time to shift to steady-state economies. Such an economy—which could be local, regional, national, or global—"aims for stable or mildly fluctuating levels in population and consumption of energy and materials. Birth rates equal death rates, and production rates equal depreciation rates," according to the Center for the Advancement of the Steady State Economy.[29]

It is also becoming increasingly clear that the GDP—the way we measure progress under our current economic systems—is outdated. Extreme weather-related events, such as flooding and storms, contribute to increases in GDP, as resources are brought in to deal with the mess. Damage done by Hurricanes Katrina and Sandy and the 2010 BP oil spill in the Gulf of Mexico added tens of billions to the GDP. If GDP growth is our highest aspiration, we should be praying for more weather catastrophes and oil spills. We deserve better indicators of societal well-being that extend beyond mere economic growth. Many economists and social scientists propose tools such as a "genuine progress indicator,"[30] which would include environmental and social factors as well as economic wealth. Some have suggested an Index of Sustainable Economic Welfare, which would take into account income inequality, environmental damage, and environmental asset depletion.[31] Bhutan has suggested measuring gross national happiness. After all, what good is a growing economy if it only benefits a few while so many suffer?

It's time we talked about a future when we can live with less and be happier, a time of greater social and economic equity with fewer stresses on the planet's ability to provide for us.

Epilogue

# WHERE DO WE GO FROM HERE?

WITH RAPIDLY ACCELERATING climate change threatening the very future of humanity, it will take nothing short of a revolution to turn things around. The degree of hardship and sacrifice that will entail depends on the determination, speed, and comprehensiveness of our actions. Employing a broad range of available solutions as quickly and as widely as possible could ensure that we build a healthier, cleaner world with greater equity.

The main solutions are those that will shift us away from fossil fuels as quickly as possible. Conserving energy is key, as is continuing to develop clean energy sources. The fastest-growing and most promising technologies are wind and solar, but geothermal, tidal, and some types of hydro are all important as well. Biofuels show some promise, as long as valuable land for food production isn't taken over to produce them.

As well as shifting away from fossil fuels, it's important to protect, restore, and, in some cases, build carbon sinks—forests, wetlands, and other green spaces that absorb and store carbon. Agricultural practices also play a big role, as agriculture is a main contributor to global warming. Reducing livestock production, enhancing soils, and focusing on local production are just some ways to reduce the impacts of growing and producing food. Personal solutions are also crucial. Reducing reliance on private automobiles, conserving energy, and reducing or eliminating animal products from our diets can all help. Stabilizing human population growth is also crucial. Giving women more rights over their own bodies, providing equal opportunity for them to participate in society, and making education and birth control widely available will help slow population growth and create numerous other benefits.

Many of the solutions require international, national, and local governance regulations and incentives. Global climate agreements, national emissions targets, infrastructure spending that encourages green development, incentives for renewable energy and energy retrofits for buildings, and trade agreements that make climate a major consideration are all necessary.

One thing is certain: If we continue to focus on building pipelines to get "product" to market, if we continue to dig up oil sands, frack, drill in deep water, and tear up the Arctic, we will face extreme hardship and sacrifice and possibly unimaginable catastrophe as nature corrects the imbalance we created with our greed, ignorance, and hubris. We've already stalled for so long that the gradual transition that might have once been possible is no longer an option—and the more we delay implementing solutions, the more upheaval we will face in shifting to cleaner energy and better agricultural practices. The seriousness and rapid pace of climate change mean that we have to get all hands on deck and use as many viable solutions as possible to arrest it. Keep in mind that greenhouse gases like

$CO_2$ stay in the atmosphere for a long time, so even if we were able to accomplish the impossible and stop emissions quickly, we'd still be locked into warming from the gases we've already pumped into the atmosphere. Had we acted more sensibly when we first knew the nature of the problem, we'd be further ahead than we are today. But we didn't, so we must consider a range of solutions, including rethinking the wasteful ways we live on this planet, and the economic systems that call for endless growth in a world with finite resources.

One group of scientists and experts, including Nicholas Stern and David Attenborough, started the Global Apollo Programme to Combat Climate Change "to accelerate the decarbonisation of the world economy through more rapid technical progress, achieved through an internationally-coordinated program of research and development over a 10-year period."[1] As the program's name suggests, we need a massive project along the lines of the American effort to get people on the moon—but much larger and on a global scale— to address "the greatest threat to global stability and prosperity." But we also need a massive social movement of people who are willing to challenge the prevailing orthodoxies and ideologies of waste, consumption, and inequality.

Climate change is both simple and complex. Simple because we know that the unnatural rise in global average temperatures is largely caused by burning fossil fuels, as well as by agricultural practices and damage to or destruction of natural features that absorb and store carbon. Complex because it's not easy to predict the various consequences and effects of interacting elements of the earth's natural systems. Everything is interconnected. This is especially true of the planet's natural cycles and the ways in which atmosphere, land, oceans, ice, and living things interact. Just as the causes are simultaneously simple and complex, so are the solutions. But there's no simple or single way to overcome a problem as large as global warming. Addressing climate change and its impacts will require a huge number of solutions, many of them complementary.

As I've pointed out in this book, many of those solutions are available, and the rapid pace of technological development means more and better solutions are being developed every day. The barriers to confronting global warming are more political and psychological than technical. Fear of change is one major psychological barrier. It's true that changing the infrastructure and the economic systems that have brought us to this point could come with sacrifices, but doing nothing will be far worse. And perhaps once people cut back on their desire and demand for disposable products and endless consumer goods, they'll find that those things aren't that important— that owning stuff brings far less joy than our connections with each other and with nature. Political barriers include the fact that politicians often think in terms of short election cycles, so they are often unwilling to take the big steps necessary to put us on the path to real change. International negotiations often get bogged down over questions of costs and responsibilities.

And there's still a lot of disagreement about the best approaches to confronting the problem, with some arguing for energy conservation and increased renewable energy use, others for more adaptation in the form of massive geoengineering schemes, and still others calling for a complete revision of our economic systems—and every imaginable combination of the above.

Powerful corporate, military, and government interests often promote big engineering solutions or large-scale options such as nuclear, as they keep power centralized and in the hands of those who already wield control. That doesn't mean we should rule out these options, but we need to look at how to transform our energy, agriculture, and economic systems in ways that best protect the climate and facilitate greater equity and justice.

Although free-market capitalism can help spark innovation in new technologies, our current economic systems demand endless growth, continued consumption, and the most profitable pathways, which often means using the cheapest fuels to extract the most money from energy and production. That's why some thinkers have suggested it's time to change the systems by which we govern production and distribution—to find a way that goes beyond outdated ideas such as free-market capitalism and authoritarian communism. As Naomi Klein writes in *This Changes Everything*, this should be seen as an opportunity, since the current economic model is "failing the vast majority of the people on the planet on multiple fronts." She adds, "Put another way, if there has ever been a moment to advance a plan to heal the planet that also heals our broken economies and shattered communities, this is it."[2]

Of course, this is what many who argue against fighting climate change fear: that addressing the problem will mean overturning current systems, which could take away some of the advantages the wealthy and privileged enjoy, often at the expense of the world's poorest and disadvantaged. But considering that about 10 percent of the world's wealthiest people are responsible for about half of global emissions, according to a 2015 Oxfam report, it's clear that adjustments are necessary.[3] Some options suggested by Klein include low-rate financial transaction taxes, closing tax havens, a small "billionaire's tax," cutting global military spending, a $50-per-tonne carbon tax in developed countries, and phasing out fossil fuel subsidies. Klein notes that much of the current ideology in developed nations is opposed to government interference in the market, but with a large-scale problem like climate change, she argues, "how can you win an argument against government intervention if the very habitability of the planet depends on intervening?"[4]

The energy question itself appears to be resolvable with conservation and renewable energy technologies, but other major contributors to climate change appear more complex. How do we get people to stop relying so much on private automobiles? Are fuel-efficient or electric vehicles the answer, or do they just keep us locked into a wasteful car culture? What about agriculture, a major source of emissions? Some argue for increased industrial methods and more genetically modified foods, whereas others say finding ways to increase local production and work with nature, using agroecological methods, would get us further.

Finding agreement on the best methods to confront global warming will not be easy, but we have little choice, and the more we study and discuss it, the closer we'll come to resolving it. But we can't just talk and study. We have to act.

We can't just sit on the sidelines and expect government and industry to tackle the problem, especially now that a climate change–denying president and party have been elected in the Unites States, a country whose climate action is critical to the whole world.

Governments move slowly at the best of times. We can't count on them to make the changes we so desperately need. It's up to us. We must be the change. We have our work cut out for us, but work we must. Perhaps this is even an opportunity, albeit one fraught with great challenges. The 2016 U.S. presidential election exposed nasty currents in U.S. society, but it also revealed a profound and rising dissatisfaction with the status quo. There's good reason for that. The gap between rich and poor has grown, globalization and changing technologies have left many people behind in an outdated economic system, racism is displayed daily on social media and television, education standards have declined, traditional media is breaking down, war and violence continue, and the effects of climate change worsen every day.

The answer isn't to throw more gas on the fire. Many Americans did that in 2016. Now, it's up to those of us who believe in a brighter future to bring the fire under control without killing the flame. Despite Donald Trump's promises to overturn what progress has been made on environmental and climate policies and initiatives, there's no stopping the wave already underway. Renewable energy investments have surpassed fossil fuel investments every year since 2010, and the gap continues to grow; American states and cities are putting a price on carbon, investing in renewable energy and in transit; electric vehicles will achieve price parity with gas vehicles by 2022; and the global movement against climate change is not going to stop.

Looking at history can give us hope. Sometimes in the midst of despair and chaos, it's difficult to notice that breakthroughs are occurring, that the world is shifting. When Archduke Franz Ferdinand of Austria was assassinated in June 1914, no one thought, "Uh-oh, the First World War is starting..." We only recognize the significance of events in the context of history.

Embracing scientific information about the warming planet and committing to avoid a catastrophic temperature increase this century creates a huge opportunity. The important hurdle is to commit to reduce emissions, because until we start, we won't know what opportunities will arise. In 1961, when President John F. Kennedy said the U.S. would get American astronauts safely to the moon and back in a decade, no one knew how they were going to do it. Amazingly, not only did they achieve the goal before the decade was over, but there were hundreds of totally unanticipated spinoffs, including laptops, cell phones, GPS, ear thermometers, and space blankets. I share with the proponents of the Global Apollo Programme the conviction that the same will happen when we commit to avoiding chaotic climate change.

With the Paris Agreement demonstrating that world leaders are taking climate change seriously, and with people at all levels of society making changes in their own lives, developing solutions, and rejecting the forces of "no," we can see that a revolution is underway.

Sometimes it takes disaster to bring about change. When citizens can no longer breathe the air, as in parts of China and Mexico, governments must find ways to address the problem. But progress is also about the better parts of human nature. We can and must speak louder than those who would keep us on a destructive path despite the overwhelming evidence that it's past time to shift course. When those of us who care about humanity and the planet's future stand up and speak out, we can make this small, blue planet and its miraculous life and natural systems a better place for all.

## Acknowledgements

THIS BOOK OWES a lot to a great many people. In particular, we are grateful to the scientists who continue to devote their lives to helping us understand this challenging problem, its consequences, and what we might do about it. The work is difficult and often depressing, and is met with a lot of negative reaction and sometimes threats from those who refuse to accept the reality of global warming. Those who spend their days researching, communicating, writing, and talking about the subject also deserve our gratitude, as do the many people around the world who are developing solutions, in areas from renewable energy to agricultural practices to conservation.

Closer to home, the many dedicated people at the David Suzuki Foundation have put a great deal of effort into this work—writing, researching, offering advice, fact-checking, and editing. Current and former members of the climate and clean energy team have helped a lot, including Kyle Aben, Ian Bruce, Tyler Bryant, Gideon Forman, and Steve Kux. Other staff have also contributed, including Karel Mayrand, Michelle Molnar, Faisal Moola, Jay Ritchlin, Aryne Sheppard, and Scott Wallace.

Gail Mainster has, as always, been invaluable with her editing skills and patience with last-minute editing requests.

We're incredibly thankful to Rob Sanders, Nancy Flight, and Shirarose Wilensky at Greystone Books, as well as the Greystone Books/David Suzuki Institute publishing committee, for this book and all the great books they publish on important topics.

We'd also like to thank staff at the many newspapers, magazines, and online publications that feature the weekly Science Matters column, from which some of this material is drawn. Readers who comment on and share the articles are also important to us. We thank the many readers who take the time to offer constructive feedback and to add to this critical conversation.

On a personal note, I'd like to thank my coauthor, David Suzuki, for his constant advice, writing, knowledge, and insight. He's one of the hardest-working people I've ever met, a truly wise elder. I'm always amazed that he's able to maintain compassion, optimism, and a great sense of humour through it all.

Lastly, I'd like to dedicate this book to my amazing son, Luc, who keeps me going even when I feel like giving up, and to all my wonderful nieces and nephews and grandnephews and all the children of the world, to whom we owe so much.

**IAN HANINGTON**

# Notes

## Introduction: Beyond Paris 2015

1. International Energy Agency. "Renewable Energy." www.iea.org/aboutus/faqs/renewableenergy.
2. Jeffrey Sachs, "By Separating Nature from Economics, We Have Walked Blindly Into Tragedy," Guardian, May 10, 2015, https://www.theguardian.com/global-development-professionals-network/2015/mar/10/jeffrey-sachs-economic-policy-climate-change.

## Chapter 1: The Science

1. "The Discovery of Global Warming," Center for History of Physics of the American Institute of Physics, www.aip.org/history/climate/index.htm.
2. Steve Graham, "John Tyndall (1820–1893)," NASA Earth Observatory, October 8, 1999, earthobservatory.nasa.gov/Features/Tyndall.
3. S.M. Enzler, "History of the Greenhouse Effect and Global Warming," Lenntech, www.lenntech.com/greenhouse-effect/global-warming-history.htm.
4. David Appell, "Behind the Hockey Stick," Scientific American, March 1, 2005, www.scientificamerican.com/article/behind-the-hockey-stick.
5. "Science Publishes New NOAA Analysis: Data Show No Recent Slowdown in Global Warming," National Oceanic and Atmospheric Association, June 4, 2015, www.noaanews.noaa.gov/stories2015/noaa-analysis-journal-science-no-slowdown-in-global-warming-in-recent-years.html.
6. John C. Fyfe, Gerald A. Meehl, Matthew H. England, Michael E. Mann, Benjamin D. Santer, Gregory M. Flato, Ed Hawkins, Nathan P. Gillett, Sang-Ping Xie, Yu Kosaka, and Neil C. Swart, "Making Sense of the Early-2000s

Warming Slowdown," *Nature Climate Change*, February 24, 2016, www.
nature.com/nclimate/journal/v6/n3/full/nclimate2938.html.

## Chapter 2: Consequences and Impacts

1.  James E. Hansen, "Climate Change Is Here—And Worse
    Than We Thought," *Washington Post*, August 3, 2012, www.
    washingtonpost.com/opinions/climate-change-is-here--and-worse-than-
    we-thought/2012/08/03/6ae604c2-dd90-11e1-8e43-4a3c4375504a_story.
    html.
2.  John Abraham, "More Evidence That Global Warming Is Intensi-
    fying Extreme Weather," *Guardian*, July 1, 2015, www.theguardian.
    com/environment/climate-consensus-97-per-cent/2015/jul/01/
    more-evidence-that-global-warming-is-intensifying-extreme-weather.
3.  Jeff Tollefson, "Severe Weather Linked More Strongly to Global
    Warming," *Nature*, June 24, 2015, www.nature.com/news/
    severe-weather-linked-more-strongly-to-global-warming-1.17828.
4.  Nathalie Schaller, et al., "Human Influence on Climate in the 2014 South-
    ern England Winter Floods and Their Impacts," *Nature Climate Change*,
    February 1, 2016, www.nature.com/nclimate/journal/vaop/ncurrent/full/
    nclimate2927.html.
5.  Barry Saxifrage, "Arctic 'Death Spiral' Leaves Climate Scientists
    Shocked and Worried," *Vancouver Observer*, September 19, 2012, www.
    vancouverobserver.com/blogs/climatesnapshot/arctic-death-spiral-leaves-
    climate-scientists-shocked-and-worried?page=0,0.
6.  John Vidal, "Arctic Expert Predicts Final Collapse of Sea Ice within
    Four Years," *Guardian*, September 17, 2012, www.theguardian.com/
    environment/2012/sep/17/arctic-collapse-sea-ice?newsfeed=true.
7.  Arctic Snow and Ice Data Center, nsidc.org/arcticseaicenews.
8.  Mark Serreze, "Why Is the Arctic So Sensitive to Climate Change and
    Why Do We Care?" NOAA, www.arctic.noaa.gov/essay_serreze.html.
9.  "The Regional Impacts of Climate Change," IPCC, www.ipcc.ch/ipccre-
    ports/sres/regional/index.php?idp=43.
10. Jim Efstathiou Jr., "Rig Grounding Revives Debate Over Shell's Arctic
    Drilling," *Bloomberg*, January 3, 2013, www.bloomberg.com/news/articles/
    2013-01-03/rig-grounding-revives-debate-over-shell-s-arctic-drilling.

11. "Arctic vs. Antarctic," National Snow and Ice Data Center, nsidc.org/cryosphere/seaice/characteristics/difference.html.

12. Brian Clark Howard, "What Antarctica's Incredible 'Growing' Icepack Really Means," *National Geographic*, November 3, 2015, news. nationalgeographic.com/2015/11/151103-antarctic-ice-growing-shrinking-glaciers-climate-change.

13. Ibid.

14. Chris Pash, "This Is Why Antarctic Seas Have Been So Resilient to Climate Change," *Business Insider*, June 3, 2016, www.businessinsider.com.au/this-is-why-antarctic-seas-have-been-so-resilient to-climate-change-2016-6.

15. "Quick Facts on Ice Sheets," National Snow and Ice Data Center, nsidc.org/cryosphere/quickfacts/icesheets.html.

16. "South Pole Is Last Place on Earth to Pass Global Warming milestone," NOAA, June 15, 2016, www.noaa.gov/south-pole-last-place-on-earth-to-pass global warming-milestone.

17. John Abraham, "New Methods Are Improving Ocean and Climate Measurements," *Guardian*, June 20, 2016, www.theguardian.com/environment/climate-consensus-97-per-cent/2016/jun/20/new-methods-are-improving-ocean-and-climate-measurements.

18. Zoe Holman, "Plastic Debris Reaches Southern Ocean, Previously Thought to Be Pristine," *Guardian*, September 27, 2012, www.theguardian.com/environment/2012/sep/27/plastic-debris-southern-ocean-pristine.

19. Alex D. Rogers, ed., "The State of the Ocean Report 2013," International Programme on the State of the Ocean, September 30, 2013, www.stateoftheocean.org/science/state-of-the-ocean-report.

20. "Latest Review of Science Reveals Ocean in Critical State," International Union for Conservation of Nature (IUCN), October 3, 2013, www.iucn.org/content/latest-review-science-reveals-ocean-critical-state.

21. Hans-O Pörtner, David Karl, et al., IPCC WGII AR5, Chapter 6: Ocean Systems, October 28, 2013, ipcc-wg2.gov/AR5/images/uploads/WGI-IAR5-Chap6_FGDall.pdf.

22. Yadigar Sekerci and Sergei Petrovskii, "Mathematical Modelling of Plankton–Oxygen Dynamics Under the Climate Change," *Bulletin of Mathematical Biology*, December 2015, link.springer.com/article/10.1007%2FS11538-015-0126-0.

23. Randy Shore, "Acidic Water Blamed for West Coast Scallop Die-Off," *Vancouver Sun*, February 24, 2014, www.vancouversun.com/life/Acidic+water+blamed+West+Coast+scallop/9550861/story.html.

24. "Ocean Acidification's Impact on Oysters and Other Shellfish," NOAA PMEL Carbon Program, www.pmel.noaa.gov/co2/story/ Ocean+Acidification%27s+impact+on+oysters+and+other+shellfish.

25. "*Science* Publishes New NOAA Analysis: Data Show No Recent Slowdown in Global Warming," NOAA, June 4, 2015, www.noaanews.noaa.gov/ stories2015/noaa-analysis-journal-science-no-slowdown-in-global-warming-in-recent-years.html.

26. Alex D. Rogers, ed., "The State of the Ocean Report 2013," International Programme on the State of the Ocean," September 30, 2013, www.stateoftheocean.org/science/state-of-the-ocean-report.

27. "Climate Change Indicators in the United States: Oceans," U.S. Environmental Protection Agency, www3.epa.gov/climatechange/science/ indicators/oceans.

28. "Sea Level Rise: Ocean Levels Are Getting Higher—Can We Do Anything About It?" *National Geographic*, ocean.nationalgeographic.com/ocean/ critical-issues-sea-level-rise.

29. "WHO Calls for Stronger Action on Climate-Related Health Risks," World Health Organization, August 27, 2014, www.who.int/mediacentre/news/ releases/2014/climate-health-risks-action/en.

30. "Climate Change and Health," World Health Organization, September 2015, www.who.int/mediacentre/factsheets/fs266/en.

31. Wendy Koch, "Climate Change Linked to More Pollen, Allergies, Asthma," *USA Today*, May 31, 2013, www.usatoday.com/story/news/nation/2013/05/ 30/climate-change-allergies-asthma/2163893/.

32. Tyler Hamilton, "Climate Change Is Wreaking Havoc on Our Mental Health, Experts Say," *Toronto Star*, February 28, 2016, www.thestar.com/ news/world/2016/02/28/climate-change-is-wreaking-havoc-on-our-mental-health-experts.html.

33. Kevin J. Coyle and Lise Van Susteren, "The Psychological Effects of Global Warming on the United States," National Wildlife Federation, February 2012, www.nwf.org/pdf/Reports/Psych_Effects_Climate_Change_ Full_3_23.pdf.

34. David R. Boyd, *The Optimistic Environmentalist: Progressing Towards a Greener Future* (Toronto: ECW Press, 2015).

35. Janneke Hille Ris Lambers, "Extinction Risks from Climate Change," *Science*, May 1, 2015, science.sciencemag.org/content/348/6234/501.

36. Helen Fields and Alanna Mitchell, "Heavy Metal Songs: Contaminated Songbirds Sing the Wrong Tunes," *Environmental Health News*, August 28, 2014, www.environmentalhealthnews.org/ehs/news/2014/aug/wingedwarnings4heavy-metal-songs.

37. "New Assessment Highlights Climate Change as Most Serious Threat to Polar Bear Survival—IUCN Red List," International Union for Conservation of Nature (IUCN), November 19, 2015, www.iucn.org/content/new-assessment-highlights-climate-change-most-serious-threat-polar-bear-survival-iucn-red.

38. Ibid.

39. "Extinction Crisis Continues Apace," IUCN, November 3, 2009, www.iucn.org/content/extinction-crisis-continues-apace.

40. John Wendle, "The Ominous Story of Syria's Climate Refugees," *Scientific American*, December 17, 2015, www.scientificamerican.com/article/ominous-story-of-syria-climate-refugees.

41. Ibid.

42. Craig Bennett, "Failure to Act on Climate Change Means an Even Bigger Refugee Crisis," *Guardian*, September 7, 2015, www.theguardian.com/environment/2015/sep/07/climate-change-global-warming-refugee-crisis.

43. "Climate Refugee," *National Geographic*, nationalgeographic.org/encyclopedia/climate-refugee.

44. Maria Sarraf, "Two Scenarios for a Hotter and Drier Arab World—And What We Can Do About It," World Bank, November 24, 2014, blogs.worldbank.org/arabvoices/two-scenarios-hotter-and-drier-arab-world-and-what-we-can-do-about-it.

45. Nikolay N. Damyanov, H. Damon Matthews, and Lawrence A. Mysak, "Observed Decreases in the Canadian Outdoor Skating Season Due to Recent Winter Warming," IOP Science, March 5, 2012, iopscience.iop.org/article/10.1088/1748-9326/7/1/014028/meta;jsessionid=5ECAABOFIE-998AF65904F431699D819C.c5.iopscience.cld.iop.org.

46. Ian Bruce, *On Thin Ice: Winter Sports and Climate Change*, David Suzuki Foundation, March 2009, davidsuzuki.org/publications/reports/2009/on-thin-ice-winter-sports-and-climate-change.

47. "Laurier Researchers Use Backyard Rinks to Track Climate Change Communications, Public Affairs & Marketing," Wilfrid Laurier University, January 8, 2013, legacy.wlu.ca/news_detail.php?grp_id=0&nws_id=10558.

## Chapter 3: Obstacles and Barriers

1. Adam Minter, "Why Living in Beijing Could Ruin Your Life," *Bloomberg*, June 19, 2014, www.bloomberg.com/view/articles/2014-06-19/why-living-in-beijing-could-ruin-your-life.

2. Elizabeth Economy, "China and Climate Change: Three Things to Watch after Paris," *Forbes*, December 15, 2015, www.forbes.com/sites/elizabetheconomy/2015/12/15/china-and-climate-change-three-things-to-watch-after-paris/#3be5797d117e.

3. "Renewable Energy Prospects: China," International Renewable Energy Agency (IRENA), November 2014, www.irena.org/menu/index.aspx?mnu=Subcat&PriMenuID=36&CatID=141&SubcatID=480.

4. "Air Pollution, Heart Disease and Stroke," Heart and Stroke Foundation of Canada Position Statement, June 2009, www.heartandstroke.com/site/apps/nlnet/content2.aspx?c=ikIQLcMWJtE&b=4869055&ct=7134879&printmode=1.

5. "Global Greenhouse Gas Emissions Data," U.S. EPA, www3.epa.gov/climatechange/ghgemissions/global.html.

6. Mengpin Ge, Johannes Friedrich, and Thomas Damassa, "6 Graphs Explain the World's Top 10 Emitters," World Resources Institute, November 25, 2014, www.wri.org/blog/2014/11/6-graphs-explain-world%E2%80%99s-top-10-emitters.

7. Jeremy Schulman, "2 GOP Candidates Have Reasonable Positions on Climate Change. They Won't Be in Tonight's Debate," *Mother Jones*, November 10, 2015, www.motherjones.com/blue-marble/2015/11/fox-business-debate-republicans-climate-change.

8. Naomi Klein, *This Changes Everything* (Toronto: Vintage Canada, 2015), 63.

9. Jonathan Watts, "Uruguay Makes Dramatic Shift to Nearly 95% Electricity from Clean Energy," *Guardian*, December 3, 2015, www.theguardian.com/

environment/2015/dec/03/uruguay-makes-dramatic-shift-to-nearly-95-clean-energy.

10. "Renewable Energy: Moving towards a Low Carbon Economy," European Commission, ec.europa.eu/energy/en/topics/renewable-energy.

11. Cassie Werber, "Three European Countries Have Already Hit Their 2020 Renewable Energy Goals," *Quartz*, March 10, 2015, qz.com/359415/three-european-countries-have-already-hit-their-2020-renewable-energy-goals.

12. "Denmark Leads the Charge in Renewable Energy," *Deutsche Welle*, May 2, 2015, www.dw.com/en/denmark-leads-the-charge-in-renewable-energy/a 17603695.

13. "Reduction of GHG Emissions From Ships," International Maritime Organization, July 25, 2014, www.imo.org/en/OurWork/Environment/PollutionPrevention/AirPollution/Documents/MEPC%2067-INF.3%20-%20Third%20IMO%20GHG%20study%202014%20-%20Final%20Report%20(Secretariat).pdf.

14. Martin Cames, Jakob Graichen, Anne Siemons, and Vanessa Cook, "Emissions Reduction Targets for International Aviation and Shipping," European Parliament, November 2015, www.europarl.europa.eu/RegData/etudes/STUD/2015/569964/IPOL_STU(2015)569964_EN.pdf.

15. Les Whittington, "Ottawa Faces $250-Million Suit over Quebec Environmental Stance," *Toronto Star*, December 26, 2012, www.thestar.com/news/canada/2012/11/15/ottawa_faces_250million_suit_over_quebec_environmental_stance.html.

16. "Lone Pine Resources Inc. v. The Government of Canada, ICSID Case No. UNCT/15/2," Italaw, 2015, www.italaw.com/cases/1606.

17. Stuart Trew, "Canadians Are Nervous about China Trade Pact. They Should Be," *iPolitics*, November 14, 2012, ipolitics.ca/2012/11/14/dnp-trew.

18. "Ontario Not Ready to Allow Fracking," CBC via the Canadian Press, November 20, 2012, www.cbc.ca/news/canada/toronto/ontario-not-ready-to-allow-fracking-1.1140920.

19. "WTO Rules Ontario Green Energy Tariff Unfair," CBC, November 19, 2012, www.cbc.ca/news/business/wto-rules-ontario-green-energy-tariff-unfair-1.1185686.

20. "Canada Joins Trans-Pacific Partnership Trade Talks," CBC, October 9, 2012, www.cbc.ca/news/politics/canada-joins-trans-pacific-partnership-trade-talks-1.1147885.

21. Naomi Klein, *This Changes Everything* (Toronto: Vintage Canada, 2015), 359.
22. Tom Faunce, "Why Australia Walked Away from Investor State Rights in Trade Deals, and Why Canada Should Too," Troy Media, November 14, 2012, www.troymedia.com/2012/11/14/why-australia-walked-away-from-investor-state-rights-in-trade-deals.
23. "Enbridge Northern Gateway Project Joint Review Panel," National Energy Board of Canada, 2013, gatewaypanel.review-examen.gc.ca/clf-nsi/dcmnt/rcmndtnsrprt/rcmndtnsrprtvlm1-eng.html.
24. Peter O'Neil, "Kinder Morgan Pipeline Application Says Oil Spills Can Have Both Negative and Positive Effects," *Vancouver Sun*, April 29, 2014, www.vancouversun.com/news/Kinder+Morgan+pipeline+application+says+spills+have+both+negative+positive+effects/9793673/story.html#ixzz30VA6zs68.
25. "Rail Industry Affirms Both Rail and Pipelines Will Move Crude Oil Safely and Reliably across North America," Association of American Railroads, January 23, 2014, www.aar.org/newsandevents/Press-Releases/Pages/Rail-Industry-Affirms-Both-Rail-and-Pipelines-Will-Move-Crude-Oil-Safely-and-Reliably-Across-North-America.aspx.
26. Marc Lee and Amanda Card, "A Green Industrial Revolution: Climate Justice, Green Jobs and Sustainable Production in Canada," Canadian Centre for Policy Alternatives, June 12, 2012, www.policyalternatives.ca/publications/reports/green-industrial-revolution.
27. "Cars and Global Warming," Union of Concerned Scientists, www.ucsusa.org/clean-vehicles/car-emissions-and-global-warming#.VI8vkOYrLYp.
28. Constantine Boussalis and Travis G. Coan, "Text-Mining the Signals of Climate Change Doubt," *Global Environmental Change*, January 2016, www.sciencedirect.com/science/article/pii/s0959378015300728.
29. Graham Readfearn, "Era of Climate Science Denial Is Not Over, Study Finds," *Guardian*, January 7, 2016, www.theguardian.com/environment/planet-oz/2016/jan/07/era-of-climate-science-denial-is-not-over-study-finds?CMP=share_btn_fb.
30. Justin Farrell, "Corporate Funding and Ideological Polarization about Climate Change," PNAS, October 12, 2015, www.pnas.org/content/113/1/92.full.
31. Abe Streep, "In Las Vegas, Climate Change Deniers Regroup, Vow to Keep Doubt Alive," *Bloomberg*, July 14, 2014, www.bloomberg.com/news/articles/

2014-07-10/in-las-vegas-climate-change-deniers-regroup-vow-to-keep-doubt-alive.

32. Rasmus E. Benestad, Dana Nuccitelli, Stephan Lewandowsky, Katharine Hayhoe, Hans Olav Hygen, Rob van Dorland, and John Cook, "Learning from Mistakes in Climate Research," *Theoretical and Applied Climatology*, August 20, 2015, link.springer.com/article/10.1007/s00704-015-1597-5.

33. Dana Nuccitelli, "Here's What Happens When You Try to Replicate Climate Contrarian Papers," *Guardian*, August 25, 2015, www.theguardian.com/environment/climate-consensus-97-per-cent/2015/aug/25/heres-what-happens-when-you-try-to-replicate-climate-contrarian-papers.

34. Bill McKibben, "Why the Energy Industry Is So Invested in Climate Change Denial," *Guardian*, February 7, 2012, www.theguardian.com/commentisfree/cifamerica/2012/feb/07/why-energy-industry-so-invested-climate-denail

35. Naomi Klein, *This Changes Everything* (Toronto: Vintage Canada, 2015), 40.

36. Fiona Harvey, "Climate Change Is Already Damaging Global Economy, Report Finds," *Guardian*, September 26, 2012, www.theguardian.com/environment/2012/sep/26/climate-change-damaging-global-economy.

37. "Climate Vulnerability Monitor: A Guide to the Cold Calculus of a Hot Planet," DARA *Impact Letters*, 2012, daraint.org/climate-vulnerability-monitor/climate-vulnerability-monitor-2012.

38. "New Study Adds Up the Benefits of Climate-Smart Development in Lives, Jobs, and GDP," World Bank, June 23, 2014, www.worldbank.org/en/news/feature/2014/06/23/study-adds-up-benefits-climate-smart-development-lives-jobs-gdp.

39. "The Bottom Line on Climate Change," *Risky Business*, riskybusiness.org.

40. Henry M. Paulson Jr., "The Coming Climate Crash: Lessons for Climate Change in the 2008 Recession," *New York Times*, June 21, 2014, www.nytimes.com/2014/06/22/opinion/sunday/lessons-for-climate-change-in-the-2008-recession.html?_r=0.

41. Justin Gillis, "Bipartisan Report Tallies High Toll on Economy From Global Warming," *New York Times*, June 24, 2014, www.nytimes.com/2014/06/24/science/report-tallies-toll-on-economy-from-global-warming.html.

42. "New Study Adds Up the Benefits of Climate-Smart Development in Lives, Jobs, and GDP," World Bank, June 23, 2014, www.worldbank.org/en/news/feature/2014/06/23/study-adds-up-benefits-climate-smart-development-lives-jobs-gdp.

43. Wendy J. Palen, Thomas D. Sisk, Maureen E. Ryan, Joseph L. Árvai, Mark Jaccard, Anne K. Salomon, Thomas Homer-Dixon, and Ken P. Lertzman, "Energy: Consider the Global Impacts of Oil Pipelines," Nature, June 25, 2014, www.nature.com/news/energy-consider-the-global-impacts-of-oil-pipelines-1.15434.

44. "New Comment in Nature on Oil Sands," Palen Lab, June 25, 2014, palen-lab.wordpress.com/oilsands.

45. Charles Bowden, Blood Orchid: An Unnatural History of America (New York: North Point Press, 2002).

46. "Justice Louis Brandeis," Brandeis University, www.brandeis.edu/legacy-fund/bio.html.

47. Canadian Index of Wellbeing, uwaterloo.ca/canadian-index-wellbeing.

48. "Frequently Asked Questions," Global Footprint Network, www.footprint-network.org/en/index.php/GFN/page/frequently_asked_questions.

49. "State of World Population 2011," United Nations Population Fund, 2011, www.unfpa.org/publications/state-world-population-2011.

## Part 2: The Solutions

1. Jeremy Lent, "False Solutions? 3 Ways to Evaluate Grand Climate Proposals," Huffington Post, March 24, 2016, www.huffingtonpost.com/jeremy-lent/false-solutions-3-ways-to_b_9534974.html.

## Chapter 4: Personal Solutions

1. Gary Gardner, "Power to the Pedals," World Watch Magazine, July/August 2010, www.worldwatch.org/node/6456.

2. "Bicycle Commuting Data," League of American Bicyclists, www.bikeleague.org/commutingdata.

3. Brian McKenzie, "Modes Less Traveled—Bicycling and Walking to Work in the United States: 2008–2012," United States Census Bureau, May 2014, www.census.gov/prod/2014pubs/acs-25.pdf.

4. Mark Hume, "Vancouver Mayor May Pay the Political Price for Bike Lanes," *Globe and Mail*, July 31, 2011, www.theglobeandmail.com/news/british-columbia/vancouver-mayor-may-pay-the-political-price-for-bike-lanes/article626103.

5. "12 Benefits of Walking," Arthritis Foundation, www.arthritis.org/living-with-arthritis/exercise/workouts/walking/wow-of-walking.php.

6. Elisabeth Rosenthal, "Across Europe, Irking Drivers Is Urban Policy," *New York Times*, June 26, 2011, www.nytimes.com/2011/06/27/science/earth/27traffic.html?_r=1.

7. Lawrence Frank, et al., "Transportation and Health in Metro Vancouver," www.myhealthmycommunity.org/Portals/0/Documents/MHMC%20Transportation%20and%20Health%20VPUBLIC%2012MAR2015.pdf.

8. Angie Schmitt, "The Koch Brothers' War on Transit," Streetsblog USA, September 25, 2014, usa.streetsblog.org/2014/09/25/the-koch-brothers-war-on-transit.

9. "Koch Industries Secretly Funding the Climate Denial Machine," Greenpeace, www.greenpeace.org/usa/global-warming/climate-deniers/koch-industries.

10. Sunny Freeman, "Koch Brothers Are Largest U.S. Lease-Holders In Oil-sands," *Huffington Post*, March 21, 2014, www.huffingtonpost.ca/2014/03/21/koch-brothers-keystone-oilsands_n_5008748.html?.

11. Angie Schmitt, "Peeking behind the Curtain of Big Oil Funded Agenda 21 Conspiracy Mongers," Streetsblog USA, July 12, 2012, usa.streetsblog.org/2012/07/12/peeking-behind-the-curtain-of-big-oil-funded-agenda-21-conspiracy-mongers.

12. Ben Adler, "The Koch brothers Just Kicked Mass Transit in the Face," *Grist*, February 3, 2015, grist.org/climate-energy/the-koch-brothers-just-kicked-mass-transit-in-the-face.

13. Ben Sichel, "The Canadian Taxpayers Federation: A Myopic Watchdog?," August 20, 2010, www.dominionpaper.ca/articles/3609.

14. Jen St. Denis, "Creator of Ethical Oil Site Will Run 'No' Campaign for Metro Vancouver Transit Referendum," *Business in Vancouver*, January 7, 2015, www.biv.com/article/2015/1/creator-ethical-oil-website-will-run-no campaign-m.

15. Susan Shaheen and Nelson Chan, "Mobility and the Sharing Economy: Impacts Synopsis," Transportation Sustainability Research Center— University of California, Berkeley, Spring 2015, innovativemobility.org/ wp-content/uploads/2015/07/Innovative-Mobility-Industry-Outlook_ SM-Spring-2015.pdf.

16. "Number of Car Sharing Users Worldwide from 2006 to 2014 (in Millions)," Statista, www.statista.com/statistics/415636/car-sharing-number-of-users-worldwide.

17. "Global Number of Car Sharing Users to Reach 650 Million by 2030," ABI Research, March 9, 2015, www.abiresearch.com/press/global-number-of-car-sharing-users-to-reach-650-mi.

18. Jennifer Langston, "Fuel Economy Improvements in US Climate Commitment on Par with 1970s Gains," *University of Washington Today*, December 15, 2015, www.washington.edu/news/2015/12/15/fuel-economy-improvements-in-us-climate-commitment-on-par-with-1970s-gains.

19. Emily Chung, "Cannabis Electric Car to Be Made in Canada," CBC, August 23, 2010, www.cbc.ca/news/technology/cannabis-electric-car-to-be-made-in-canada-1.903115.

20. HumanCar Inc., www.humancar.com.

21. "Average Household Electricity Use around the World," Shrink That Footprint, shrinkthatfootprint.com/average-household-electricity-consumption.

22. "Always-On Inactive Devices May Devour $19 Billion Worth of Electricity Annually," NRDC, May 6, 2015, www.nrdc.org/media/2015/150506-0.

23. John Schueler, "Are Energy Vampires Sucking You Dry?," U.S. Department of Energy, October 29, 2015, energy.gov/articles/are-energy-vampires-sucking-you-dry.

24. "Green Power Markets," U.S. Department of Energy, apps3.eere.energy. gov/greenpower/markets/pricing.shtml?page=0.

25. "Climate and Life Cycle of Stuff," U.S. EPA, 2009, www3.epa.gov/ climatechange/climate-change-waste.

26. Roni A. Neff, Marie L. Spiker, and Patricia L. Truant, "Wasted Food: U.S. Consumers' Reported Awareness, Attitudes, and Behaviors," *PLoS One*, June 10, 2015, journals.plos.org/plosone/article?id=10.1371%2Fjournal. pone.0127881.

27. "Infant and Young Child Feeding," World Health Organization, January 2016, www.who.int/mediacentre/factsheets/fs342/en.

28. "Food Wastage Footprint: Impacts on Natural Resources," Food and Agriculture Organization of the United Nations (FAO), 2013, www.fao.org/docrep/018/i3347e/i3347e.pdf.

29. Kip Andersen and Keegan Kuhn, Cowspiracy: The Sustainability Secret, A.U.M. Films and Media, 2014, www.cowspiracy.com.

30. "Unlocking the Energy in Foods," Science Learning, September 5, 2011, sciencelearn.org.nz/Contexts/Digestion-Chemistry/Science-Ideas-and-Concepts/Unlocking-the-energy-in-foods.

31. "Rearing Cattle Produces More Greenhouse Gases Than Driving Cars, UN Report Warns," UN News Centre, November 29, 2006, www.un.org/apps/news/story.asp?newsID=20772#.VI9BluYrKBI.

32. Chris Mooney, "The Hidden Driver of Climate Change That We Too Often Ignore," Washington Post, March 9, 2016, www.washingtonpost.com/news/energy-environment/wp/2016/03/09/the-hidden-driver-of-climate-change-that-we-too-often-ignore.

33. Bryan Walsh, "The Triple Whopper Environmental Impact of Global Meat Production," Time, December 16, 2013, science.time.com/2013/12/16/the-triple-whopper-environmental-impact-of-global-meat-production.

34. Carl Zimmer, "Inuit Study Adds Twist to Omega-3 Fatty Acids' Health Story," New York Times, September 17, 2015, www.nytimes.com/2015/09/22/science/inuit-study-adds-twist-to-omega-3-fatty-acids-health-story.html?_r=1.

35. "World Agriculture: Towards 2015/2030—Ch. 3.3 Livestock Commodities," FAO, www.fao.org/docrep/005/y4252e/y4252e05b.htm.

36. Fiona Harvey, "Eat Less Meat to Avoid Dangerous Global Warming, Scientists Say," Guardian, March 21, 2016, www.theguardian.com/environment/2016/mar/21/eat-less-meat-vegetarianism-dangerous-global-warming.

37. Michael Pollan, "Unhappy Meals," New York Times Magazine, January 28, 2007, michaelpollan.com/articles-archive/unhappy-meals.

38. Sara Cullen, "BoG Votes against Divestment Following Release of CAMSR Report," McGill Tribune," March 30, 2016, www.mcgilltribune.com/news/bog-votes-divestment-following-release-camsr-report-290316.

39. Karel Mayrand, "Why I Will Hand Back My degree to McGill University," David Suzuki Foundation, April 1, 2016, www.davidsuzuki.org/blogs/ panther-lounge/2016/04/why-i-will-hand-back-my-degree-to-mcgill-university.

40. "Genus Fossil Free Is Pioneering a Profitable New Type of Impact Investing," Genus Capital Management Inc., www.genusfossilfree.com/ more-profitable.

41. Melanie Mattauch, "In the Space of Just 10 Weeks...," Fossil Free—350.org, December 2, 2015, gofossilfree.org/in-the-space-of-just-10-weeks.

42. "Your Roadmap to Divestment," Fossil Free—350.org, gofossilfree.org/ your-roadmap-to-divestment.

43. Hunter Lovins, "Life after Divestment: How to Spend the Money Saved from Fossil Fuel Investments," *Guardian*, April 13, 2015, www.theguardian. com/sustainable-business/2015/apr/13/divestment-colleges-universities-finance-fossil-fuels-investments.

## Chapter 5: Agricultural Solutions

1. Stephen Russell, "Everything You Need to Know About Agricultural Emissions," World Resources Institute, May 29, 2014, www.wri.org/ blog/2014/05/everything-you-need-know-about-agricultural-emissions.

2. "Global Greenhouse Gas Emissions Data," U.S. EPA, www3.epa.gov/climat-echange/ghgemissions/global.html.

3. James Owen, "Farming Claims Almost Half Earth's Land, New Maps Show," *National Geographic*, December 9, 2005, news.nationalgeographic. com/news/2005/12/1209_051209_crops_map.html.

4. "What Is Happening to Agrobiodiversity?," UN Food and Agriculture Organization, www.fao.org/docrep/007/y5609e/y5609e02.htm.

5. Tom Parrett, "GMO Scientists Could Save The World From Hunger, If We Let Them," *Newsweek*, May 21, 2015, www.newsweek.com/2015/05/29/ gmo-scientists-could-save-world-hunger-if-we-let-them-334119.html.

6. Joel Achenbach, "107 Nobel Laureates Sign Letter Blasting Greenpeace over GMOs," *Washington Post*, June 30, 2016, www.washingtonpost.com/ news/speaking-of-science/wp/2016/06/29/more-than-100-nobel-laureates-take-on-greenpeace-over-gmo-stance/?postshare=7871467217813235&tid= ss_tw.

7.  Gerry Everding, "Genetically Modified Golden Rice Falls Short on Lifesaving Promises," *The Source*, Washington State University in St. Louis, June 2, 2016, source.wustl.edu/2016/06/genetically-modified-golden-rice-falls-short-lifesaving-promises.

8.  Rachel Carson, *Silent Spring* (New York: Houghton Mifflin, 1962).

9.  Salvatore Ceccarelli, "The Centrality of Seed: Building Agricultural Resilience through Plant Breeding," *Independent Science News*, February 29, 2016, www.independentsciencenews.org/un-sustainable-farming/the-centrality-of-seed-building-agricultural-resilience-through-plant-breeding

10. Ibid.

11. Jeremy Lent, "False Solutions? 3 Ways to Evaluate Grand Climate Proposals," *Huffington Post*, May 24, 2016, www.huffingtonpost.com/jeremy-lent/false-solutions-3-ways-to-b_9534974.html.

12. Miguel Altieri, "Agroecology: Principles and Strategies for Designing Sustainable Farming Systems," University of California, Berkeley, 2000, www.agroeco.org/doc/new_docs/Agroeco_principles.pdf.

13. "Regenerative Organic Agriculture and Climate Change," Rodale Institute, April 17, 2014, rodaleinstitute.org/regenerative-organic-agriculture-and-climate-change.

14. "UN Expert Makes Case for Ecological Farming Practices to Boost Food Production," UN News Centre, March 8, 2011, www.un.org/apps/news/story.asp?NewsID=37704#.V19MD-YiKBo.

15. Ibid.

16. Ute Scheub, Haiko Pieplow, Hans-Peter Schmidt, and Kathleen Draper, *Terra Preta: How the World's Most Fertile Soil Can Help Reverse Climate Change and Reduce World Hunger* (Vancouver: Greystone Books/David Suzuki Institute, 2016), xiv.

17. Mark Hertsgaard, "As Uses of Biochar Expand, Climate Benefits Still Uncertain," Yale Environment 360, January 21, 2014, e360.yale.edu/feature/as_uses_of_biochar_expand_climate_benefits_still_uncertain/2730/

18. Ute Scheub, et al., *Terra Preta: How the World's Most Fertile Soil Can Help Reverse Climate Change and Reduce World Hunger* (Vancouver: Greystone Books/David Suzuki Institute, 2016), 6.

19. Ibid., xv.

20. Tom Miles, "Amazon's Mysterious Black Earth: Soil Found along Region Riverbanks; Rich in Nutrients, Stores More Carbon," BioEnergy Lists, January 20, 2007, www.biochar.bioenergylists.org/forestsorg.

21. Susan S. Lang, "Cornell Biogeochemist Shows How Reproducing the Amazon's Black Soil Could Increase Fertility and Reduce Global Warming," Cornell Chronicle, February 18, 2006, www.news.cornell.edu/stories/2006/02/amazonian-black-soil-could-improve-soils-reduce-global-warming.

22. Dawit Solomon, Johannes Lehmann, James Angus, and James Fairhead, "Indigenous African Soil Enrichment as a Climate-Smart Sustainable Agriculture Alternative," Frontiers in Ecology and the Environment, February 2016, www.researchgate.net/publication/296619044_Indigenous_African_soil_enrichment_as_a_climate-smart_sustainable_agriculture_alternative.

23. "700 Year-Old African Soil Technique Could Help to Mitigate Climate Change," University of Sussex, News & Events, www.sussex.ac.uk/newsandevents/fertile-soil.

24. Ute Scheub, et al., Terra Preta: How the World's Most Fertile Soil Can Help Reverse Climate Change and Reduce World Hunger (Vancouver: Greystone Books/David Suzuki Institute, 2016), 91.

25. Ibid., xvi.

26. Ibid., 76.

27. "World's Population Increasingly Urban with More Than Half Living in Urban Areas," United Nations, July 10, 2014, www.un.org/en/development/desa/news/population/world-urbanization-prospects-2014.html.

28. Peter Ladner, The Urban Food Revolution: Changing the Way We Feed Cities (Gabriola Island, B.C.: New Society Publishers, 2011).

29. Raychel Santo, Anne Palmer, and Brent Kim, "Vacant Lots to Vibrant Plots: A Review of the Benefits and Limitations of Urban Agriculture," Johns Hopkins Center for a Livable Future, May 2016, www.jhsph.edu/research/centers-and-institutes/johns-hopkins-center-for-a-livable-future/_pdf/research/clf_reports/urban-ag-literature-review.pdf.

30. Lani Furbank, "Johns Hopkins Report Analyzes the Benefits and Limitations of Urban Agriculture," Food Tank, June 5, 2016, foodtank.com/news/2016/06/johns-hopkins-report-analyzes-the-benefits-and-limitations-of-urban-ag.

31. Raychel Santo, Anne Palmer, and Brent Kim, "Vacant Lots to Vibrant Plots: A Review of the Benefits and Limitations of Urban Agriculture," Johns Hopkins Center for a Livable Future, May 2016, www.jhsph.edu/research/centers-and-institutes/johns-hopkins-center-for-a-livable-future/_pdf/research/clf_reports/urban-ag-literature-review.pdf.

32. Ibid.

33. Peter Ladner, *The Urban Food Revolution: Changing the Way We Feed Cities* (Gabriola Island, B.C.: New Society Publishers, 2011).

34. David Suzuki Foundation, Homegrown National Park, homegrown. projexity.com.

## Chapter 6: Technological Solutions

1. Ralph D. Torrie, Tyler Bryant, Mitchell Beer, Blake Anderson, Dale Marshall, Ryan Kadowaki, and Johanne Whitmore, *An Inventory of Low-Carbon Energy for Canada*, David Suzuki Foundation, March 2013, www.davidsuzuki.org/publications/reports/2013/an-inventory-of-low-carbon-energy-for-canada.

2. Rob Jordan, "Stanford Researcher Maps out an Alternative Energy Future for New York," *Stanford News*, March 12, 2013, news.stanford.edu/news/2013/march/new-york-energy-031213.html.

3. Elisabeth Rosenthal, "Life After Oil and Gas," *New York Times*, March 23, 2013, www.nytimes.com/2013/03/24/sunday-review/life-after-oil-and-gas.html?pagewanted=all&_r=4&amp.

4. Ibid.

5. Ralph D. Torrie, *Power Shift: Cool Solutions to Global Warming*, David Suzuki Foundation, 2000, davidsuzuki.org/publications/reports/2000/power-shiftcool-solutions-to-global-warming.

6. "Can Renewables Provide Baseload Power?" Skeptical Science, www.skepticalscience.com/renewable-energy-baseload-power.htm.

7. David Roberts, "More on Ramping down Baseload Power and Ramping up Storage," *Grist*, March 30, 2012, grist.org/energy-policy/more-on-ramping-down-baseload-power-and-ramping-up-storage.

8. Michael Mariotte, "The Archaic Nature of Baseload Power—Or Why Electricity Will become Like Long Distance," *GreenWorld*, August 20, 2015.

9. David Mills, "Busting the Baseload Power Myth," ABC *Science*, December 2, 2010, www.abc.net.au/science/articles/2010/12/02/3081889.htm.

10. David Roberts, "Why Germany Is Phasing out Nuclear Power," *Grist*, March 23, 2012, grist.org/renewable-energy/why-germany-is-phasing-out-nuclear-power.

11. Karel Beckman, "Steve Holliday, CEO National Grid: 'The Idea of Large Power Stations for Baseload Is Outdated,'" *Energy Post*, September 11, 2015, www.energypost.eu/interview-steve-holliday-ceo-national-grid-idea-large-power-stations-baseload-power-outdated.

12. David Roberts, "Want More Wind and Solar? We'll Need to Get Rid of Outdated Grid Rules," *Vox*, July 23, 2015, www.vox.com/2015/7/23/9020019/energy-markets-wind-and-solar.

13. Ibid.

14. Ibid.

15. Michael Mariotte, "The Archaic Nature of Baseload Power—Or Why Electricity Will become Like Long Distance," *GreenWorld*, August 20, 2015.

16. "What You Need to Know about Energy," National Academy of Sciences, www.nap.edu/read/12204/#intro.

17. "Solar Energy," *National Geographic*, environment.nationalgeographic.com/environment/global-warming/solar-power-profile.

18. David Chandler, "What's the Future of Solar Power?" World Economic Forum, April 2, 2015, www.weforum.org/agenda/2015/04/whats-the-future-of-solar-power.

19. "The Future of Solar Energy," MIT Energy Initiative, May 5, 2015, mitei.mit.edu/futureofsolar.

20. Renee Cho, "Where Is Solar Power Headed?" Phys.org, July 22, 2015, phys.org/news/2015-07-solar-power.html.

21. Nancy W. Stauffer, "Study Assesses Solar Photovoltaic Technologies," Phys.org, December 17, 2015, phys.org/news/2015-12-solar-photovoltaic-technologies.html#jCp.

22. Lynn Yarris, "Major Advance in Artificial Photosynthesis Poses Win/Win for the Environment," Berkeley Lab, April 16, 2015, newscenter.lbl.gov/2015/04/16/major-advance-in-artificial-photosynthesis.

23. "SolaRoad Generates More Power Than Expected," CBC/*Associated Press*, May 12, 2015, www.cbc.ca/news/technology/solaroad-generates-more-power-than-expected-1.3069371.

24. Karel Beckman, "Steve Holliday, CEO National Grid: 'The Idea of Large Power Stations for Baseload Is Outdated,'" *Energy Post*, September 11, 2015,

www.energypost.eu/interview-steve-holliday-ceo-national-grid-idea-large-power-stations-baseload-power-outdated.

25. Arthur Neslen, "Morocco Poised to become a Solar Superpower with Launch of Desert Mega-Project," *Guardian*, October 26, 2015, www.theguardian.com/environment/2015/oct/26/morocco-poised-to-become-a-solar-superpower-with-launch-of-desert-mega-project.

26. Nicholas Brown, "Coldest Parts of Earth Have the Best Solar (PV) Potential, Study Finds," CleanTechnica, October 23, 2011, cleantechnica.com/2011/10/23/coldest-parts-of-earth-have-the-best solar pv-potential-study-finds.

27. "Wind Energy," BP Global, www.bp.com/en/global/corporate/energy-economics/statistical-review-of-world-energy/renewable-energy/wind-energy.html.

28. "Wind in Numbers," Global Wind Energy Council, www.gwec.net/global-figures/wind-in-numbers.

29. "Birds and Buildings," Ontario Nature, www.ontarionature.org/protect/campaigns/birds_and_buildings.php.

30. Martin Harper, "Facing up to Inconvenient Truths," Royal Society for the Protection of Birds, April 7, 2013, www.rspb.org.uk/community/ourwork/b/martinharper/archive/2013/04/07/facing-up-to-inconvenient-truths.aspx.

31. Alan Neuhauser, "Pecking Order: Energy's Toll on Birds," *U.S. News & World Report*, August 22, 2014, www.usnews.com/news/blogs/data-mine/2014/08/22/pecking-order-energys-toll-on-birds.

32. "Wind Turbine Noise and Health Study: Summary of Results," Health Canada, October 30, 2014, www.hc-sc.gc.ca/ewh-semt/noise-bruit/turbine-eoliennes/summary-resume-eng.php.

33. "Wind Turbine Health Impact Study: Report of Independent Expert Panel," Massachusetts Department of Environmental Protection and Massachusetts Department of Public Health, January 2012, www.mass.gov/eea/docs/dep/energy/wind/turbine-impact-study.pdf.

34. Simon Chapman, Alexis St. George, Karen Waller, and Vince Cakic, "The Pattern of Complaints about Australian Wind Farms Does Not Match the Establishment and Distribution of Turbines: Support for the Psychogenic, 'Communicated Disease' Hypothesis," *PLoS One*, October 16, 2013, journals.plos.org/plosone/article?id=10.1371/journal.pone.0076584.

35. Fiona Crichton, George Dodd, Gian Schmid, Greg Gamble, and Keith J. Petrie, "Can Expectations Produce Symptoms from Infrasound Associated

with Wind Turbines?" *Health Psychology*, April 2014, psycnet.apa.org/psycinfo/2013-07740-001.

36. "Breakdown of Electricity Generation by Energy Source," The Shift Project, www.tsp-data-portal.org/Breakdown-of-Electricity-Generation-by-Energy-Source#tspQvChart.

37. "Research Reveals Dramatic Growth Of Global Hydropower Expected This Decade," *WaterWorld*, October 24, 2014, www.waterworld.com/articles/2014/10/global-boom-in-hydropower-expected-this-decade.html.

38. "Hydroelectric Power Water Use," U.S. Geological Survey, water.usgs.gov/edu/wuhy.html.

39. "Hydropower," Center for Climate and Energy Solutions, www.c2es.org/technology/factsheet/hydropower.

40. Julia Pyper, "World's Dams Unprepared for Climate Change Conditions," *Scientific American*, September 16, 2011, www.scientificamerican.com/article/worlds-dams-unprepared-for-climate-change.

41. John H. Matthews, Bart A.J. Wickel, and Sarah Freeman, "Converging Currents in Climate-Relevant Conservation: Water, Infrastructure, and Institutions," *PLoS Biology*, September 6, 2011, journals.plos.org/plosbiology/article?id=10.1371%2Fjournal.pbio.1001159.

42. Eric Holthaus, "Hot Dam," June 1, 2015, www.slate.com/articles/business/moneybox/2015/06/the_future_of_hydroelectricity_it_s_not_good.html.

43. Joshua Robertson, "Australia's Answer to Tesla: Indigenous Firm AllGrid Shines in Solar Battery Industry," *Guardian*, December 26, 2015, www.theguardian.com/environment/2015/dec/26/australias-answer-to-tesla-indigenous-firm-allgrid-shines-in-solar-battery-industry.

44. Umair Irfan, "Battery Storage Needed to Expand Renewable Energy," *Scientific American*, February 13, 2015, www.scientificamerican.com/article/battery-storage-needed-to-expand-renewable-energy.

45. *E-Storage: Shifting from Cost to Value 2016*, World Energy Council, 2016, www.worldenergy.org/publications/2016/e-storage-shifting-from-cost-to-value-2016.

46. Ibid.

47. Bjorn Carey, "Stanford Engineers Develop State-by-State Plan to Convert U.S. to 100% Clean, Renewable Energy by 2050," *Stanford News*, June 8, 2015, news.stanford.edu/pr/2015/pr-50states-renewable-energy-060815.html.

48. Ibid.

49. Richard Blackwell, "Ontario Seeks Wind, Solar Energy Storage Options," *Globe and Mail*, August 3, 2014, www.theglobeandmail.com/report-on-business/industry-news/energy-and-resources/ontario-picks-contenders-for-wind-solar-energy-storage/article19901932.

50. "How Energy Storage Works," Union of Concerned Scientists, www.ucsusa.org/clean-energy/how-energy-storage-works#.VI9naeYrkBI.

51. "Renewable Integration Benefits," Energy Storage Association, energystorage.org/energy-storage/energy-storage-benefits/benefit-categories/renewable-integration-benefits.

52. Robert F. Service, "New Type of 'Flow Battery' Can Store 10 Times the Energy of the Next Best Device," *Science*, November 27, 2015, www.sciencemag.org/news/2015/11/new-type-flow-battery-can-store-10-times-energy-next-best-device.

53. "Organic Mega Flow Battery Promises Breakthrough for Renewable Energy," Harvard John A. Paulson School of Engineering and Applied Sciences, January 8, 2014, https://www.seas.harvard.edu/news/2014/01/organic-mega-flow-battery-promises-breakthrough-for-renewable-energy.

54. Tesla Motors, "Powerwall," www45.teslamotors.com/en_CA/powerwall.

55. "US Electricity Could Be Powered Mostly by the Sun and Wind by 2030: Rapid, Affordable Energy Transformation Possible," *Science Daily*, January 25, 2016, www.sciencedaily.com/releases/2016/01/160125114231.htm.

56. Alexander E. MacDonald, Christopher T. M. Clack, Anneliese Alexander, Adam Dunbar, James Wilczak, and Yuanfu Xie, "Future Cost-Competitive Electricity Systems and Their Impact on US $CO_2$ Emissions," *Nature Climate Change*, January 25, 2016, www.nature.com/nclimate/journal/v6/n5/full/nclimate2921.html.

57. Emma Gilchrist, "Geothermal Offers Cheaper, Cleaner Alternative to Site C Dam: New Report," DeSmog Canada, November 25, 2014, www.desmog.ca/2014/11/25/geothermal-offers-cheaper-cleaner-alternative-site-c-dam-new-report.

58. Raphael Lopoukhine, "Top 5 Reasons Why Geothermal Power Is Nowhere in Canada," DeSmog Canada, February 27, 2014, www.desmog.ca/2014/02/26/top-5-reasons-why-geothermal-power-nowhere-canada.

59. "Geothermal Energy," *National Geographic*, environment.nationalgeographic.com/environment/global-warming/geothermal-profile.

60. Bjorn Carey, "Stanford Engineers Develop State-by-State Plan to Convert U.S. to 100% Clean, Renewable Energy by 2050," *Stanford News*, June 8, 2015, news.stanford.edu/pr/2015/pr-50states-renewable-energy-060815.html.

61. Kennedy Maize, "Geothermal Energy: Is New Technology Resetting the Agenda?," *Power*, October 1, 2015, www.powermag.com/geothermal-energy-is-new-technology-resetting-the-agenda.

62. Peter Dockrill, "Scientists Are Using Satellites to Find Untapped Sources of Geothermal Energy under Cities," *Science Alert*, January 15, 2016, www.sciencealert.com/scientists-are-using-satellites-to-find-untapped-sources-of-geothermal-energy-under-cities.

63. Wendy Koch, "Creator of 5-Hour Energy Wants to Power the World's Homes—With Bikes," *National Geographic*, October 6, 2015, news.nationalgeographic.com/energy/2015/10/151006-energy-drink-billionaire-wants-to-power-homes-with-bikes/.

64. Ibid.

65. "Biofuels," *National Geographic*, environment.nationalgeographic.com/environment/global-warming/biofuel-profile.

66. "Biofuels Overview," Center for Climate and Energy Solutions, www.c2es.org/technology/overview/biofuels.

67. Tiffany Stecker, "Biofuels Might Hold Back Progress Combating Climate Change," *Scientific American*, March 31, 2014, www.scientificamerican.com/article/biofuels-might-hold-back-progress-combating-climate-change.

68. Andrew Steer and Craig Hanson, "Biofuels Are Not a Green Alternative to Fossil Fuels," *Guardian*, January 29, 2015, www.theguardian.com/environment/2015/jan/29/biofuels-are-not-the-green-alternative-to-fossil-fuels-they-are-sold-as.

69. "The Current Status of Biofuels in the European Union, Their Environmental Impacts and Future Prospects," European Academies Science Advisory Council, December 2012, www.easac.eu/fileadmin/PDF_s/reports_statements/Easac_12_Biofuels_Complete.pdf.

70. "Ethanol Tanks," *The Economist*, October 22, 2009, www.economist.com/node/14710469.

71. David Biello, "Grass Makes Better Ethanol than Corn Does," *Scientific American*, January 8, 2008, www.scientificamerican.com/article/grass-makes-better-ethanol-than-corn.

72. Dan Krotz, "From Soil Microbe to Super-Efficient Biofuel Factory?," Berkeley Lab, May 3, 2012, newscenter.lbl.gov/2012/05/03/electrofuel.

73. "Should Toronto Burn Its Trash for Energy?," City News, August 14, 2013, www.citynews.ca/2013/08/14/should-toronto-burn-its-trash-for-energy.

74. "Canadian Waste Management Statistics Show Recycling on the Rise," Waste Management World, January 7, 2011, waste-management-world.com/a/canadian-waste-management-statistics-show-recycling-on-the-rise.

75. Nate Seltenrich, "Is Incineration Holding Back Recycling?," Guardian, August 29, 2013, www.theguardian.com/environment/2013/aug/29/incineration-recycling-europe-debate-trash.

76. "Negative Impacts of Incineration-Based Waste-to-Energy Technology," Alternative Energy News, September 2008, www.alternative-energy-news.info/negative-impacts-waste-to-energy.

77. Marc Lee, Sue Maxwell, Ruth Legg, and William Rees, "Closing the Loop: Reducing Greenhouse Gas Emissions through Zero Waste in BC," Canadian Centre for Policy Alternatives, March 28, 2013, www.policyalternatives.ca/publications/reports/closing-loop.

78. Simon L. Lewis, "Tropical Forests Grab Carbon," Nature, February 19, 2009, www.nature.com/nature/journal/v457/n7232/edsumm/e090219-07.html.

79. Donald L. Hey, Deanna L. Montgomery, Laura S. Urban, Tony Prato, Andrew Forbes, Mark Martell, Judy Pollack, Yoyi Steele, and Ric Zarwell, "Flood Damage Reduction in the Upper Mississippi River Basin: An Ecological Alternative," Wetlands Initiative, August 6, 2004, static1.squarespace.com/static/567070822399a343227dd9c4/t/568d6213c647ad-1e518d2b07/1452106259180/flood_damage_reduction_in_umrb.pdf.

80. Erika Thorkelson, "30 Years of Calgary Flood Warnings Fell on Deaf Ears," DeSmog Canada, June 25, 2013, www.desmog.ca/2013/06/25/30-years-calgary-flood-warnings-fell-deaf-ears.

81. "VCS Approves 'Coastal Blue Carbon' as New International Carbon Trading Category," Restore America's Estuaries, www.estuaries.org/vcs-approves-qcoastal-blue-carbonq-as-new-international-carbon-trading-category.

82. "What Is Natural Capital?" David Suzuki Foundation, www.davidsuzuki.org/issues/wildlife-habitat/projects/natural-capital/what-is-natural-capital.

83. Bowker Creek Initiative, Capital Regional District (Victoria), www.crd.bc.ca/bowker-creek-initiative.

84. "New York City," Information Center for the Environment, UC Davis, ice. ucdavis.edu/node/133.

85. Doug Koplow, "Nuclear Power: Still Not Viable without Subsidies," Union of Concerned Scientists, February 2011, www.ucsusa.org/sites/default/ files/legacy/assets/documents/nuclear_power/nuclear_subsidies_report. pdf.

86. Jeremy Lent, "False Solutions? 3 Ways to Evaluate Grand Climate Proposals," *Huffington Post*, March 24, 2016, www.huffingtonpost.com/jeremy-lent/ false-solutions-3-ways-to_b_9534974.html.

87. Andrew C. Revkin, "To Those Influencing Environmental Policy but Opposed to Nuclear Power," *New York Times*, November 3, 2013, dotearth. blogs.nytimes.com/2013/11/03/to-those-influencing-environmental-poli- cy-but-opposed-to-nuclear-power/?_r=0.

88. Justin Gillis, "In Search of Energy Miracles," *New York Times*, March 11, 2013, www.nytimes.com/2013/03/12/science/in-search-of-energy-miracles. html?_r=1&am.

89. "Thorium," World Nuclear Association, September 2015, www.world-nu- clear.org/information-library/current-and-future-generation/thorium.aspx.

90. Phil McKenna, "Is the 'Superfuel' Thorium Riskier Than We Thought?," *Popular Mechanics*, December 5, 2012, www.popularmechanics.com/science/ energy/a11907/is-the-superfuel-thorium-riskier-than-we-thought-14821644.

91. Eifion Rees, "Don't Believe the Spin on Thorium Being a Greener Nuclear Option," *Guardian*, June 23, 2011, www.theguardian.com/ environment/2011/jun/23/thorium-nuclear-uranium.

92. Duncan Geere, "UK Report Says Thorium Nuke Power Potential 'Overstated,'" *Wired*, September 17, 2012, www.wired.com/2012/09/ thorium-report.

93. Matthew Bramley, "Is Natural Gas a Climate Change Solution for Canada?" David Suzuki Foundation and Pembina Institute, July 2011, www. davidsuzuki.org/publications/reports/2011/is-natural-gas-a-climate- change-solution-for-canada.

94. David Biello, "Fracking's Biggest Problem May Be What to Do with Wastewater," *Scientific American*, June 22, 2012, blogs.scientificamerican.com/ observations/frackings-biggest-problem-may-be-what-to-do-with- wastewater.

95. Monika Freyman, "Hydraulic Fracturing & Water Stress: Water Demand by the Numbers," Ceres, February 2014, www.ceres.org/issues/water/shale-energy/shale-and-water-maps/hydraulic-fracturing-water-stress-water-demand-by-the-numbers.

96. Steve Hockensmith, "Why State's Water Woes Could Be Just Beginning," Berkeley News, January 21, 2014, news.berkeley.edu/2014/01/21/states-water-woes.

97. Terry Reith and Briar Stewart, "Do Fracking Activities Cause Earthquakes? Seismologists and the State of Oklahoma Say Yes," CBC, April 28, 2016, www.cbc.ca/news/canada/edmonton/fracking-debate-earthquakes-oklahoma-1.3554275.

98. Lauren Krugel, "Fracking Causing Minor Earthquakes in B.C.: Regulator," Canadian Press/Financial Post, September 6, 2012, business.financialpost.com/news/energy/fracking-causing-minor-earthquakes-in-b-c-regulator?_lsa=0ad2-75cd.

99. Bill McKibben, "Why Not Frack?," New York Review of Books, March 8, 2012, www.nybooks.com/articles/2012/03/08/why-not-frack/?pagination=false.

100. David Biello, "Can Controversial Ocean Iron Fertilization Save Salmon?," Scientific American, October 24, 2012, www.scientificamerican.com/article/fertilizing-ocean-with-iron-to-save-salmon-and-earn-money.

101. Aaron Strong, Sallie Chisholm, Charles Miller, and John Cullen, "Ocean Fertilization: Time to Move On," Nature, September 17, 2009, www.nature.com/nature/journal/v461/n7262/full/461347a.html.

102. Jessica Marshall, "Ocean Geoengineering May Prove Lethal," ABC Science, March 16, 2010, www.abc.net.au/science/articles/2010/03/16/2846973.htm.

103. "The Global Status of CCS," Global CCS Institute, 2015, status.globalccsinstitute.com.

104. Ian Austen, "Technology to Make Clean Energy From Coal Is Stumbling in Practice," New York Times, March 29, 2016, www.nytimes.com/2016/03/30/business/energy-environment/technology-to-make-clean-energy-from-coal-is-stumbling-in-practice.html?_r=0.

105. John Shepherd, "Geoengineering the Climate: Science, Governance and Uncertainty," Royal Society, September 1, 2009, royalsociety.org/topics-policy/publications/2009/geoengineering-climate.

106. Michael Zürn and Stefan Schäfer, "The Paradox of Climate Engineering," *Global Policy*, July 29, 2013, onlinelibrary.wiley.com/doi/10.1111/gpol.12004/ abstract.

107. Jeremy Lent, "False Solutions? 3 Ways to Evaluate Grand Climate Proposals," *Huffington Post*, March 24, 2016, www.huffingtonpost.com/jeremy-lent/false-solutions-3-ways-to_b_9534974.html.

108. Naomi Klein, *This Changes Everything* (Toronto: Vintage Canada, 2015), 283.

## Chapter 7: Institutional Solutions

1. Kate Connolly, "G7 Leaders Agree to Phase out Fossil Fuel Use by End of Century," *Guardian*, June 8, 2015, www.theguardian.com/world/2015/jun/08/g7-leaders-agree-phase-out-fossil-fuel-use-end-of-century.

2. "10 Reasons Scientists Are Calling for a Moratorium of the Oil Sands," www.oilsandsmoratorium.org.

3. Wendy J. Palen, et al., "Energy: Consider the Global Impacts of Oil Pipelines," *Nature*, June 25, 2014, www.nature.com/news/energy-consider-the-global-impacts-of-oil-pipelines-1.15434.

4. John S. Daniel, Guus J.M. Velders, et al., "Scientific Assessment of Ozone Depletion: 2010," NOAA, www.esrl.noaa.gov/csd/assessments/ozone/2010/summary/ch5.html.

5. "Ozone Treaty Anniversary Gifts Big Birthday Present to Human Health and Combating of Climate Change," UN Environment Programme, September 16, 2009, www.unep.org/environmentalgovernance/News/PressRelease/tabid/427/language/en-US/Default.aspx?DocumentID=596&ArticleID=6305&Lang=en.

6. Naomi Oreskes and Erik Conway, *Merchants of Doubt: How a Handful of Scientists Obscured the Truth on Issues from Tobacco Smoke to Global Warming* (New York: Bloomsbury Press, 2011).

7. Beatrice Lugger, "Interview with Sherwood Rowland—Climate Change, Ozone, Misleading Campaigns," Lindau Nobel Laureate Meetings, July 19, 2010, www.lindau-nobel.org/interview-with-sherwood-rowland-climate-change-ozone-misleading-campaigns.

8. Jeff Rubin and David Suzuki, "Canada Needs a Single Price on Carbon," *Toronto Star*, September 2, 2015, www.thestar.com/opinion/commentary/2015/09/02/canada-needs-a-single-price-on-carbon.html.

9. "Sulfur Dioxide," U.S. EPA, www3.epa.gov/airtrends/sulfur.html.

10. Paul Krugman, "Building a Green Economy," New York Times, April 7, 2010, www.nytimes.com/2010/04/11/magazine/11Economy-t.html?_r=1.

11. "Steady State Economy—Definition," Center for the Advancement of the Steady State Economy, steadystate.org/discover/definition.

12. Christopher Barrington-Leigh, Bronwen Tucker, and Joaquin Kritz Lara, "The Short-Run Household, Industrial, and Labour Impacts of the Quebec Carbon Market," Canadian Public Policy, December 2015, www.utpjournals. press/doi/full/10.3138/cpp.2015-015.

13. Lara Dahan, Marion Afriat, and Katherine Rittenhouse, "California: An Emissions Trading Case Study," Institute for Climate Economics, April 2015, www.i4ce.org/download/california-an-emissions-trading-case-study.

14. Jeff Rubin and David Suzuki, "Canada Needs a Single Price on Carbon," Toronto Star, September 2, 2015, www.thestar.com/opinion/commentary/2015/09/02/canada-needs-a-single-price-on-carbon.html.

15. Allie Goldstein, "Voluntary Carbon Market Stalls, But Buyers See Silver Lining," Ecosystem Marketplace, June 26, 2014, www.ecosystemmarketplace. com/articles/voluntary-carbon-market-stalls-but-buyers-see-silver-lining.

16. Dimitri Pescia, Patrick Graichen, Mara Marthe Kleiner, and David Jacobs, "Understanding the Energiewende," Agora Energiewende, October 2015, www.agora-energiewende.de/fileadmin/Projekte/2015/Understanding_ the_EW/Agora_Understanding_the_Energiewende.pdf.

17. Ibid.

18. Gideon Forman, "What Can Canada Learn from Germany's Energy Transition?," David Suzuki Foundation, January 13, 2016, davidsuzuki.org/ blogs/climate-blog/2016/01/what-can-canada-learn-from-germanys-energy-transition.

19. Dimitri Pescia, et al., "Understanding the Energiewende," Agora Energiewende, October 2015, www.agora-energiewende.de/fileadmin/ Projekte/2015/Understanding_the_EW/Agora_Understanding_the_Energiewende.pdf.

20. Robert Kunzig, "Germany Could Be a Model for How We'll Get Power in the Future," National Geographic, October 15, 2015, ngm.nationalgeographic. com/2015/11/climate-change/germany-renewable-energy-revolution-text.

21. "How Feed-In Tariffs Maximize the Benefits of Renewable Energy," Pembina Institute, www.pembina.org/reports/feed-in-tariffs-factsheet.pdf.

22. Naomi Klein, *This Changes Everything* (Toronto: Vintage Canada, 2015), 24.

23. "New Study Adds up the Benefits of Climate-Smart Development in Lives, Jobs, and GDP," World Bank, June 23, 2014, www.worldbank.org/en/news/feature/2014/06/23/study-adds-up-benefits-climate-smart-development-lives-jobs-gdp.

24. Justin Gillis, "Bipartisan Report Tallies High Toll on Economy From Global Warming," *New York Times*, June 24, 2014, www.nytimes.com/2014/06/24/science/report-tallies-toll-on-economy-from-global-warming.html.

25. Ibid.

26. "Climate Change and Health: Fact Sheet No. 266," World Health Organization, September 2015, www.who.int/mediacentre/factsheets/fs266/en.

27. Richard Wilkinson and Kate Pickett, *The Spirit Level: Why More Equal Societies Almost Always Do Better* (London: Allen Lane, 2009).

28. "What Is the Stern Review?," *Guardian*, February 15, 2011, www.theguardian.com/environment/2011/feb/15/stern-review.

29. Center for the Advancement of the Steady State Economy, "Definition," steadystate.org/discover/definition/.

30. Bob Willard, "5 Reasons Why a GPI Should Replace the GDP," Sustainability Advantage, March 8, 2011, sustainabilityadvantage.com/2011/03/08/5-reasons-why-a-gpi-should-replace-the-gdp.

31. "What Is an Indicator of Sustainability?," Sustainable Measures, www.sustainablemeasures.com/node/89.

**Epilogue: Where Do We Go from Here?**

1. Damian Carrington, "Global Apollo Programme Seeks to Make Clean Energy Cheaper Than Coal," *Guardian*, June 2, 2015, www.theguardian.com/environment/2015/jun/02/apollo-programme-for-clean-energy-needed-to-tackle-climate-change.

2. Naomi Klein, *This Changes Everything* (Toronto: Vintage Canada, 2015), 155.

3. "Extreme Carbon Inequality," Oxfam, December 2, 2015, www.oxfam.org/en/research/extreme-carbon-inequality.

4. Naomi Klein, *This Changes Everything* (Toronto: Vintage Canada, 2015), 43.

# Index

Abraham, John, 46

Agora Energiewende, 237

agriculture: and greenhouse gas emissions, 139–42, 146–47; industrial, 5, 40, 148–52; livestock, 139–42, 248; and runoff, 46, 47, 50, 149, 155; urban, 149, 158–61; use of fertilizer in, 149, 155, 157, 194

agroecology, 148, 152–54; dark earth, 148, 154–57, 160

albedo feedback, 41

AllGrid Energy, 183–84, 188

Altieri, Miguel, 153

American Institute of Physics, 19–20, 21

Americans for Prosperity, 123

ammonia. See under greenhouse gases

Antarctica: melting of ice in, xii, 44–46, 49; ozone layer, 228; plastic fragments in, 47

the Arctic: and ice ages, 19, 21; melting of ice in, 24, 29, 36, 37, 40–44, 49; ozone layer, 228; wildlife, 58–59

Arrhenius, Svante, 19, 107

Arthritis Foundation, 120

Asimov, Isaac, 23

Attenborough, David, 249

Australia, 44, 56, 75, 79–80, 101, 159–60, 179, 183–84

automobiles. See cars

Aziz, Michael J., 188

Baliunas, Sallie, 26

Bangladesh, 61

bats, 177, 178

B.C. Oil and Gas Commission, 213

Berlin Social Science Research Center, 221

Bhargava, Manoj, 191

bicycles. See cycling

biochar. See dark earth

biodiversity, 47, 148, 153, 155, 225, 241; loss of, 149–50, 201. See also extinction of wildlife

biofuels, 192–96, 247

# DAVID
## SUZUKI
## INSTITUTE

**THE DAVID SUZUKI INSTITUTE** is a non-profit organization founded in 2010 to stimulate debate and action on environmental issues. The Institute and the David Suzuki Foundation both work to advance awareness of environmental issues important to all Canadians.

We invite you to support the activities of the Institute. For more information please contact us at:

David Suzuki Institute
219–2211 West 4th Avenue
Vancouver, BC, Canada V6K 4S2
info@davidsuzukiinstitute.org
604-742-2899
www.davidsuzukiinstitute.org

Cheques can be made payable to The David Suzuki Institute.